ENGINEERING FOR SUSTAINABILITY

ENGINEERING
FOR
SUSTAINABILITY

A PRACTICAL GUIDE FOR
SUSTAINABLE DESIGN

GERALD JONKER AND JAN HARMSEN
University of Groningen, Groningen, The Netherlands

AMSTERDAM • BOSTON • HEIDELBERG • LONDON • NEW YORK • OXFORD
PARIS • SAN DIEGO • SAN FRANCISCO • SINGAPORE • SYDNEY • TOKYO

Elsevier
Radarweg 29, PO Box 211, 1000 AE Amsterdam, The Netherlands
The Boulevard, Langford Lane, Kidlington, Oxford OX5 1GB, UK

First edition 2012

Notice
No responsibility is assumed by the publisher for any injury and/or damage to
persons or property as a matter of products liability, negligence or otherwise, or
from any use or operation of any methods, products, instructions or ideas
contained in the material herein. Because of rapid advances in the medical
sciences, in particular, independent verification of diagnoses and drug dosages
should be made

British Library Cataloguing in Publication Data
A catalogue record for this book is available from the British Library

Library of Congress Cataloging-in-Publication Data
A catalog record for this book is available from the Library of Congress

For information on all Elsevier publications
visit our web site at elsevierdirect.com

ISBN: 978-0-444-53846-8

Working together to grow
libraries in developing countries

www.elsevier.com | www.bookaid.org | www.sabre.org

ELSEVIER   BOOK AID   Sabre Foundation
            International

To my family and my friends, for their support and inspiration, especially Esther, Werry, Tiny, and Elly, and our children Daan and Koen, as the future generation.

Gerald Jonker

To my parents Herman Harmsen and Lien Ebbers, my wife Mineke Spruijt, our children Joost, Maaike, Wouter, Julia and Emily, and our grandchildren: Thomas, Prisca, Maria, Anna, Lise, Sep, Minke, Nathalie.

Jan Harmsen

Contents

Preface

Sustainable development is now one of the main drivers in many businesses. For sometime in the past, it was only part of the official policy of the companies, but now business targets are set for making a significant contribution to sustainable development. To meet these targets, manufacturing processes and products have to change. These changes are to be made by engineers and engineers make these changes via designing new processes and new products. The design is the language vehicle by which the constructors can make the new process and the means by which the product manufacturers can make the product. Sustainable design is thus the key step in meeting the sustainable business targets.

In our industrial experience, we have discovered however that it is not so easy to make designs that contribute to sustainable development. The reasons are that the concept of sustainable development has wide-ranging implications, totally different dimensions relating to social, environmental, and economics, hence a long-term view has to be taken and total life cycle effects have to be taken into account. In fact, the sustainable design has to be embedded in the social, economic, and environmental context in such a way that it addresses the needs of people so that the local society prospers, no harm is done to the local and global environment, and that the companies involved make long-lasting profits.

As engineers love structure, the book is highly structured. Chapter 1 contains an introduction to sustainable design, discussing the system approach and an appropriate description of sustainability. Chapter 2 provides a step-by-step method for the actual design process in which for each step problem definition, synthesis, analysis, and evaluation, guiding principles and hints are provided, all in view of obtaining a sustainable design result. Chapters 2 and 3 describe the four context levels, planet, society, business, and engineers, so that the engineer can quickly identify which context levels are relevant for certain aspects of the specific design to be made. Chapter 4 provides methods and tools for the design steps. Chapter 5 contains proven methods for teaching and acquiring sustainable design competences.

We have written this book to aid the professional designers in industry and to aid students and professors. The information in this book has been tested in many years of teaching sustainable design in academic and industrial settings. The book is meant for all engineering disciplines and for designing any artifact. To that end, only generic design steps, guidelines, and methods are provided. We expect that professionals outside the engineering disciplines find the book useful for making plans or policies, because these activities can also be seen as design activities.

Finally, the book will also be useful for those who quickly want to acquaint themselves with the main aspects of sustainable development.

ACKNOWLEDGMENTS

We gratefully thank the reviewers for their numerous and very fruitful comments on the manuscript. From industry, Sebastiaan Frijling MSc (Senior IT Management Consultant at IBM), Dr. Ir. Michael Kuczynski (Vice President Research & Technology at DSM Fibre Intermediates and Director at DSM Research), Prof Jos Keurentjes (Technology Director AkzoNobel Industrial Chemicals) and Dr Joost Demmink MSc (Manager Technology ESD-SIC), and from the academic side, Ass.Prof. Michael Narodoslawsky (Associate Professor for Chemical Engineering, TU Graz, Austria), Prof. Ton Schoot Uiterkamp (Em Chair Environmental Sciences, University of Groningen) and Dolf Mulder M.A. (Mulder Consultancy on Ethics), and the students Ernst Erik Boer, Pieter Pesie, and Jarno Pons.

About the Authors

Gerald Jonker teaches Industrial Engineering and Management at the University of Groningen, and is an experienced university teacher of many (chemical) engineering courses at BSc, MSc and PhD level.

Jan Harmsen had thirty-three years of industrial experience at Shell with positions in: exploratory research, development, process design, and chemicals manufacturing. He is now director of Harmsen Consultancy BV. He is also a part-time Leonardo da Vinci Professor Sustainable Chemical Technology at the University of Groningen, financially supported by Shell Research Foundation.

CHAPTER

1

Introduction

On becoming a 21st Century Engineer: everything changes, everything is connected, engineering and engineers have never mattered more [1]. This book is on the professional and operational part of making a sustainable design, with practical applications, directed to all engineering disciplines. Before discussing design concepts and practical tools on sustainability, this chapter first explores sustainability, its interpretation in terms of definitions, the role of system thinking, and the scientific background (Industrial Ecology). Defining sustainability introduces the anthropocentric approach, the precautionary principle and system thinking. The latter also functions as an introduction of the Chapter (3) on Context: sustainability described in four context levels, Planet– Society–Business–Engineer. Finally a more practical and business related approach is discussed, Triple P, People, Planet, Profit/Prosperity.

Engineering for Sustainability,
DOI: 10.1016/B978-0-444-53846-8.00001-9

1

ROLE OF ENGINEERS

The future is now: it is not a matter of 'If', or 'when', but on 'how' to incorporate sustainability [2,3]. Leading business journals report papers entitled like 'There is no alternative to sustainable development' [4], see also Box 1.1. Sustainability, is convincingly stated as the next [3,5] and

---

BOX 1.1

SUSTAINABILITY AND BUSINESS

"The leading global companies of 2020 will be those that provide goods and services and reach new customers in ways that address the world's major challenges — including poverty, climate change, resource depletion, globalization, and demographic shifts". "The business is seriously incorporating corporate social responsibility." Conclusions from a committee of the World Business Council on Sustainable Development (first quote) and the second one of the leading marketers, Harvard Professor, Micheal Porter [80].

---

necessary [2] revolution. Innovation trajectories are observed to be increasingly linked to sustainability [6,4]. But what is 'sustainability'? Sustainability can be approached from many different perspectives, varying from North to South throughout the world, and from governmental regulations to market considerations. Also, professions each can have their share in working on sustainability.

Our focus is on the engineering profession. In 21st century most engineers are expected to be central in sustainability, as S.D. Sheppard states that engineering and engineers have never mattered more [1]. Engineers are working everywhere in the society, but most of their work is in the industry being the producer of products, energy, and, having an increasing share in the last decades, services. Therefore, engineering as a profession is in this respect broadly defined as being a practitioner in a technological surrounding, working on constructing artifacts in production plants, or the service industry. Their work increasingly incorporates sustainability. A first trend can be seen that throughout the past centuries, industry gradually moved from primarily focused on profit, to better labor conditions and the past decades, care of the environment or planet [2,7]. Globalization, increased transparency by better communication inspires and forces industry to subsequent steps toward sustainability, resulting in that engineers have a central role in developing and creating a sustainable world.

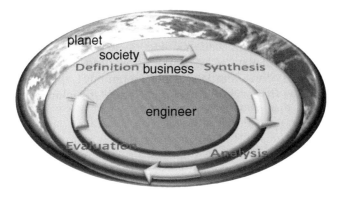

FIGURE 1.1   Schematic overview of making a sustainable design.

Figure 1.1, schematically shows the process of making a sustainable design: centered on the engineer's design assignment, the design cycle (green arrows) incorporates the outer circles representing business, society, and planet. Industry increasingly incorporates processes from outside their fences in their business strategy. A typical example is the branch of chemical producers. Companies as DSM, Akzo Nobel and DuPont are in the middle of the supply chain, producing intermediate industrial materials. Increasingly, the branch of chemical companies is getting involved in the whole supply chain, and review the impact of their products, also to the up front part and to the end user phase. Service industries have connections to many companies from different branches and via their information and knowledge providing services can have even stronger impact on the whole supply chain, because they can change the behavior of employees and their companies. Incorporating external processes usually introduce non-technical items such as, ethical considerations, awareness and insight in global trends in business and society [8]. Engineers have never mattered more in a sustainable world, in which they are challenged to incorporate the additional requirements and conditions in making a sustainable design.

EXPLORING SUSTAINABILITY

In order to properly describe what a sustainable design is, a starting point is to explore the term sustainability first. Further, despite the many different interpretations of sustainability, for application in a sustainable design a generic basic set of features of sustainability are described. An important aspect of sustainability is that it requires insight in the interdependency of subsystems, as of industrial production sites, societal trends, and environmental requirements. This system thinking implicates a holistic approach in making a sustainable design. Also, for engineers, Industrial Ecology, as the academic and research part of sustainability applied on an industry is of importance. And finally, after these more theoretical considerations, the commonly applied Triple P approach appears to be a good working metaphor in making a sustainable design.

Defining Sustainability

In the Western world, the definition is usually interpreted as sustaining the current state of welfare, closely related to environment: resources depletion, waste handling, recycling, and preservation of biodiversity [9]. However, on a global scale, the interpretation of sustainability usually is different, as it is based on the founding book 'Our Common Future', from the UN committee working in the mid 80s, chaired by Gro Harlem Brundtland [10]. Brundtland formulated personally as a compromise, a practical applicable definition of sustainability, found in the underlined text in Box 1.2. Sustainability is described as a broad term on a very

---

BOX 1.2

BRUNDTLAND DEFINITION

Sustainable development is development that meets the needs of the present without compromising the ability of future generations to meet their own needs. It contains with it two key concepts:

- The concept of 'needs', in particular the essential needs of the world's poor, to which overriding priority should be given; and
- The idea of limitations imposed by the state of technology and social organization on the environment's ability to meet present and future trends [10].

---

abstract level, and on an anthropogenic basis, thus with the human development being central. As an extension of the Brundtland definition the Triple P; People, Planet and Profit definition was later developed and in the UN Assembly of 2002 modified to People, Planet and Prosperity (further on in this chapter a more detailed description is provided). The largest part of the book 'Our Common Future' is on these three dimensions with a strong focus on the essential needs of the world's poor, closely related to inequity. The main critic on the Brundtland definition is that it is too anthropocentric, neglecting the ecological view [11]. This critic is counteracted by adding the precaution principle.

The precaution principle concentrates on the environment as it is a better-safe-than-sorry principle that advocates the reduction of inputs into the environment of substances, especially where there is a reason to believe that harmful effects are likely to occur [12]. At United Nations Conference on Environment and Development (UNCED) in Rio de Janeiro the (non-binding) Rio Declaration 1992 was agreed with the following text: (Principle 15) "In order to protect the environment, the precautionary approach shall be widely applied by States according to their capabilities. Where there are threats of serious or irreversible damage, lack of full scientific certainty shall not be used as a reason for postponing cost-effective measures to prevent environmental degradation."

A more practical interpretation of sustainability reveals a more deepening of what is actually meant by making a sustainable design. First of all, sustainability can be looked upon a so-called "contested concept". This term indicates that, as similar to 'peace', everybody intuitively understands it's meaning, however, in practice, the inter-pretation of the term may be different e.g., within a country, groups may interpret their situation differently: what is called peace by one group can be felt as being suppressed by other groups. A similar effect is occurring with sustainability: in fact the large majority is in favor for it, but the practical application is different [13]. An example is the food industry: food produced in a more sustainable way is usually judged as being more desirable, which obviously should be calculated for in prizes. However being at the mall, consumers are attracted by prizes first, and the various ways of marketing, e.g., referring to feeling well [14]. What in a Western country may be called sustainable (e.g., energy reduction), is no issue at all in developing countries.

A suitable description of a sustainable design thus requires social, cultural, and economic aspects (the anthropocentric view of Brundtland) and the precaution principle related to the environment. An example is "Sustainability in the end is a means of conducting environmentally sound economic activity for socially desirable outcomes" [15], p. 61. Socially desirable (the 'needs' of Brundtland, see above) incorporates assessment on social aspects. Human activities should not endanger global dynamics, resource availability, and the resilience of the ecosys-tems in a way that can cause problems to the self-sustainability of the current population or of future generations [16], p. 6. Important feature of sustainability is that it takes worldwide and long-term perspectives into account, emphasized by a more poetic description by John Ehrenfeld's [17]: "Sustainability is the possibility that humans and other forms of life will flourish on the planet forever".

Concluding, a sustainable design is described by: *Without compromising usual design criteria such as costs, appearance and quality, all environmental impacts of a design outcome throughout the complete life cycle should be taken into consideration, including social and economic well-being, such as ethics and the environment* [18].

System Thinking

To work on sustainability requires insight in the Human−Environment System (HES). A system is a set of interacting units with relationships among them [19], p. 16. In terms of the HES, this implies that concentrated on the human behavior, the understanding of the drivers of environmental awareness gives insight in the conflicting drivers which prevent sustainable behavior [16], p. 6. The complete picture of what today is known of the HES is very extensive and interesting, but also rather complex, demanding a thorough knowledge of interrelationships between society and environment [16,19]. For making a sustainable design, first insights and awareness of the HES enables the engineer however to explore relevant relationships.

To cope with complexity of the many relationships in the HES, hierarchy is applied to provide a structure in social systems [16], p. 418. The Miller's levels of generalized living systems range from cell, organ, individual, group, organization, society to supranational level [16], p. 414, [19]. In this book, a similar hierarchy is applied: the engineer working as an individual or in a group, in organizations, in society, and at a supranational level. Organizations can be rather broadly defined, as an assembly of people who are jointly planning, coordinating purposeful action and having a formal membership [16], p. 419. In the view of an engineer working in a job setting, organizations are here specified to business: a profit based organization, active on a market. The society level includes institutions which are under governmental control. Finally the supranational level is here shortly denoted as planet, but includes organizations as the United Nations. All four *context levels*, which will be further explored in Chapter 3, are adding to incorporate the influences of the outside (or context) to the design assignment.

Within the hierarchy of the HES, description of interconnections of industrial system and natural world (or Planet) enlightens the situation which we face today. The planet basically includes living, and from the perspective of a human time reference, regenerative and non-regenerative resources. Regenerative resources can sustain human activities indefinitely, as long as humans do not "harvest" them more rapidly than they replenish themselves. Non-regenerative resources, as mining, oil production, can only be depleted. The industrial system produces waste, which damages the ability of nature to replenish resources. Finally, the industrial system is interwoven with a larger social system of communities, families, schools, and culture, thereby providing jobs, and welfare, but also causing pressure (stress, inequity) [2], pp. 23–24. This simple picture represents a finely woven network between society, industry and the natural world, in which the work field of an engineer on the job, only interferes in a very specified subsystem. In order to make a sustainable design, only knowledge is required on the subsystem representing the context of the design assignment.

Practical applications of system thinking lead to criteria for making a sustainable design. Analysis on a global scale, such as [16,19–21] give insight to the background of these criteria. Box 1.3 provides the design criteria of The Natural Step, which is one of the most widely known sets of

---

BOX 1.3

THE NATURAL STEP

The Natural Step (TNS) formulates a practical set of four design criteria to transform debate into constructive discussion. The criteria all directly relates to conditions necessary for a sustainable society. They are to reduce and eventually eliminate our contribution to

- the build up of materials taken from the Earth's crust
- the build up of synthetic substances produced by society
- ongoing physical degradation of nature
- conditions that undermine people's ability to meet their basic needs [2], p. 382.

Source: http://www.naturalstep.org/the-system-conditions (accessed October 8, 2011)

---

principles and criteria for sustainability. Mimicking cycles in the natural world, material cycling systems are proposed, based on the postulate Waste Equals Food [7,22], using the nature's rules to build sustainable profits [22].

System thinking with the combined contribution of social, environmental and economic aspects of the design implies an holistic approach [13,23]: the combination of the elements in a system has more impact than their summation could predict. Being designers, engineers define their own subsystem, within they face the complexity of an holistic approach which states that all elements of a system (in this case the defined subsystem) cannot be separately studied, but only in their combination. Otherwise stated, a sustainable design limits the flexibility of the designer [15], p. 108, that is, probably more solutions are rejected beforehand because they do not meet requirements connected with sustainability. Also, as sustainability is focusing on the future, a serious uncertainty in these limitations is introduced because of the unpredictable long-term implications of outcomes. Current design standards usually are based on proven best practices, which may not be robust on the long term [15], p. 108. Although obviously, these limitations can be turned upside down: there are many challenges for engineers!

Experience from daily practice and from class shows that broad thinking is easier said than done, for e.g., in a casus, a motivated group of students investigated the choice of materials with a Quick Life Cycle Analysis (LCA). The group used an advanced program for determining the impact of various materials for building fisherman's houses in Sri Lanka, after the tsunami. They seriously defended that the front door should be made of mahogany wood, a very durable, locally available but also a very expensive material. From the viewpoint of LCA, it was an obvious choice, but from an holistic point of view, including local economics, it could be argued that it would be better to sell mahogany wood and use lower quality wood for the fisherman's houses. Box 1.4 summarizes this observation in requirements for engineers, see e.g., the

---

BOX 1.4

AN HOLISTIC APPROACH IN TERMS OF EDUCATION

The complexity that sustainability adds to the system of interest can be illustrated by challenges related to sustainability issues in educating engineers [24]:

- consider sustainability in all engineering decisions,
- account for social aspects,
- account for the natural environment,
- keep up-to-date,
- focus on process of acquiring solutions rather than endpoint, and
- encourage diversity within the profession.

---

fifth point: what is the effect of the intermediate steps in processes? Shortly, system thinking with an holistic approach requires a broad, dynamic and process oriented view on design assignments [24], related to continuous improvement.

Industrial Ecology

Industrial Ecology can be seen as the scientific part of sustainability as it studies the metabolism of industry, as a part of ecology, and in terms and in line with ecology [25]. In this respect, the industry is looked upon as an organism, in which parts interact resembling biological organisms. The resemblance is found in that resource flows, both in biological and industrial organisms, are transformed, e.g., in ecosystem engineering plankton blooms warm surface waters, and in industrial engineering, urban areas alter water flows [25], p. 71. The observation and starting point of IE is that the modern industrial sectors are closely integrated, as a network of technology. Represented as a tree of technology, materials and products fulfill unique niches, just as in biological symbioses [25], p. 73. Consequently, IE applies the biological toolkit to model and analyze the industrial sectors and interrelationships among them.

Ecological systems provide many insights and tools to analyze industry, such as metabolism, but also emphasize the many relationships between elements of the system. IE therefore not only concentrates on the material energy flows, but also on the information flows and interactions between system parts, such as government, society and industry; hence, it also incorporates social ecology. A strong driver in IE is that in nature, there is no such thing as waste. All materials of one cycle can be used in other cycles, e.g., manure of animal acts as fertilizer for plants, carbon dioxide is released by living creatures, but form the 'oxygen' of plants. In this respect, books as Cradle to Cradle [7], which focus on the material flow, and Earth Inc [22], on the managerial share of the life cycle, are closely connected to IE.

The steps to analyze the industrial connections consistof defining first the system of interest, after which models represent the structure or operation of an object or a system [25], p. 326. Systems may be urban ecology, or water ecology, or rather specific, the mining and usage of nickel. The system should be properly defined, to distinguish the human (driving) factors, and the environmental implications in transforming energy and materials. The scale obviously determines the amount of interrelationships in the model, which may be become complex, even on a relatively small scale. An example is the case of modeling flows of materials and money for biowaste and biowaste transformation for Canton Zurich, Switzerland. It shows the extensive material and money flow incorporated [25], p. 334. An important objective of IE is to understand by modeling these interrelationships and to research how to minimize waste and improve material and energy usage.

The knowledge provided by the analysis of IE will be of help to obtain a full description of the context of the design and make a proper assessment of the design result. IE therefore may be looked upon as a scientific foundation of the industrial oriented sustainable development. IE defines principles for Design for Environment and Sustainability, as in the well know 12 principles of green chemistry or for eco-innovation [6]. Also, the importance and usage of LCA is extensively investigated by IE. For a specific design assignment, elements of IE may be very useful to apply, but in many cases a full understanding and knowledge of all fields of IE is not necessary to make a sustainable concept design. In the final design stages however all aspects of the design and its context should be taken into account.

Triple P

After the more theoretical consideration of sustainability, system thinking, and the role of Industrial Ecology, the concept of Triple P provides a very practical application of sustainability. The financial and business world adopted the concept sustainability, but defined it further by introducing three essential dimensions of Sustainable Development namely: Social, Ecological and Economic dimensions [26,27]. This in turn was transferred into the Triple P bottom line: People, Planet, and Profit, by Elkington [28]. Soon after, a number of companies started using it (e.g., Shell [29]). According to the triple bottom line concept, equal weight should be given in corporate activities to:

- "People", the social consequences of its actions
- "Planet", the ecological consequences
- "Profit", the economic profitability of companies (being the source of "Prosperity")

The main point is that the "bottom line" of an organization is not only an economic-financial one — an organization is responsible to its social and ecological environment as well. From this "Triple P" perspective, an organization that considers a strategy of sustainability must find a balance between economic goals and goals with regard to the social and ecological environment.

Because of the Triple P success in industry, it also became popular with governments. In the United Nations World conference on sustainable development in Johannesburg, South Africa, 2002, the Triple P description was adopted and modified. The term Profit was changed into Prosperity. To quote:

> "We, the representatives of the peoples of the world, assembled at the World Summit on Sustainable Development in Johannesburg, South Africa, from 2 to 4 September 2002, reaffirm our commitment to sustainable development. We commit ourselves to act together, united by a common determination to save our planet, promote human development and achieve universal prosperity and peace"[30].

To conclude, the challenge of "Sustainability is the possibility that humans and other forms of life will flourish on the planet forever" is made more explicit in looking at system thinking and the role of Industrial Ecology. The extended design circle of Fig. 1.1 to the four context levels in terms of sustainability can be practically applied by evaluating the social, economic and environmental aspects of the design assignment, summarized as Triple P.

OVERVIEW OF MAKING A SUSTAINABLE DESIGN

Making sustainable designs is the central objective of this book. Figure 1.2 shows the structure of the book, in terms of the design cycle. The (small) left column contains the elements of the design cycle: problem definition, synthesis, analysis and evaluation. The (large) right column has two areas: at the left side general elements of any design are listed. The area at the right shows the specific methods and tools relevant for a *sustainable* design. The topics of the next two sections accordingly are General Applicable Design Steps and Design Steps Specific for a Sustainable Design, respectively. Subsequently, the book focuses on sustainable design, but can also be used for designing in general.

General Applicable Design Steps

In general human beings have a natural tendency and an ability to design when they have to solve a complex problem [31]. The method provided here is therefore meant to stimulate and facilitate this ability. It is based on long experiences of the authors with design and teaching design both in an industrial and in an academic setting. This book treats design as a problem solving activity using a proven stepwise method of problem definition, solution synthesis, analysis, evaluation and reporting.

The stepwise design approach is very fruitful for most types of design, varying from industrial oriented designs [32] to political and industrial policy design.

Figure 1.2 also shows feedback loops from analysis to synthesis and to problem definition. This is because during the analysis it may become clear that the problem is not sufficiently defined and also that the solution is not complete. In the problem definition step already preliminary solutions may come to mind. Hence, it is clear that the designing is not a simple linear logical procedure. The provided stepwise method is mainly helpful in providing hints and guiding principles in a structured way and helps inexperienced designers along their way to solutions.

Each engineering discipline has different wordings for the design steps, but the general sequence is the same. In the IT engineering discipline for instance the following steps are used: model business, analyze requirements, design, build, test, deploy. It is easy to see that business modeling and requirements analysis belong to: Define the problem, that design is the same as synthesis and that build and test belong to analyze.

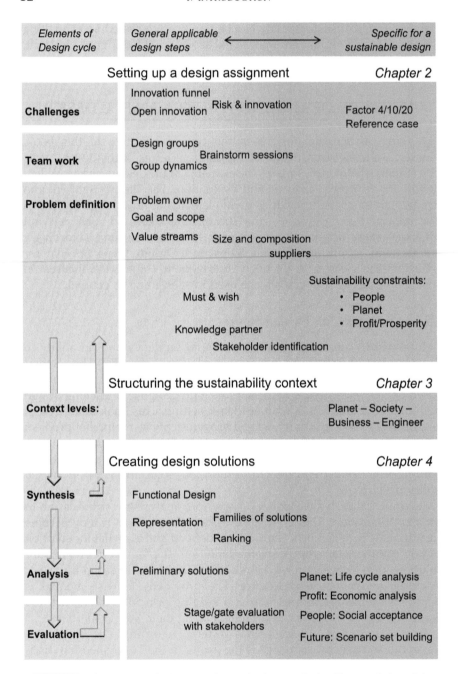

FIGURE 1.2   Sustainable design cycle as a leading guide for Chapters 2, 3, and 4.

Only the evaluation step with stakeholders is missing. But this step is essential in sustainable design and therefore should always be present.

Forming the design group is also an important part of the design. Balanced small teams consisting of members with complementary characters and skills are the best. In most cases, the design group will be a subgroup of the total research or development team. The total R&D team will be much larger. Chapter 2 provides some methods to design such a design group both for courses and for industrial purposes.

Design Steps Specific for a Sustainable Design

The right side of Fig. 1.2 contains Triple P based drivers and constraints in making a sustainable design. Hints, suggestions, and tools are provided to successfully incorporate the context of sustainability in the design assignment. As will be advocated in Chapter 2, setting up a design assignment, sustainability especially determines the first stage of the design cycle, in the problem definition. Therefore, the challenges (and innovation) are discussed, with respect to radical innovations as symbolized by the so-called Factor 4 (or 10/20) improvement, as will be discussed in Chapter 2. The last step of Chapter 2 is to define the constraints on sustainability in terms of Triple P.

To enrich the imagination of the designer regarding sustainability, Chapter 3 provides an overview of the context on sustainability. The overview aims to give fundamentals on sustainability on the four context levels, enabling the designer to explore the specific area of interest with respect to the design assignment. Sustainability contains many temporal and locally oriented projects and subjects. Chapter 3 gives insights which are, regardless time and location, generic and illustrative for sustainability.

Having set the problem statement (Chapter 2), with generic knowledge on sustainability (Chapter 3), Chapter 4 discusses the three design steps directed to create and evaluate design solutions. Regarding sustainability, four specific design tools enables to quickly analyze Triple P in a future view: Quick Scan Life Cycle Analysis (Planet), Rapid Economic Analysis (Profit), Rapid Social Acceptance (People) and Scenario Set Building (Future). These methods are directed to a conceptual design, in which possibilities are explored in making a sustainable design.

Finally, Chapter 5 is on acquiring a sustainable development mindset and methods for design by learning and teaching.

Setting up a Design Assignment

With a sustainable design, a design is meant that meets sustainable development goals and is embedded in the social, environmental and business context as described in Chapters 1 and 3. These sustainable development goals are higher and the constraints are much tighter than the conventional design goals and constraints. To sum up: The difference of a sustainable design compared to a conventional design is two-fold:

1. Sustainable Design has much wider reaching sustainable development goals and constraints

Engineering for Sustainability,
DOI: 10.1016/B978-0-444-53846-8.00002-0

2. The sustainable designing process is different because a holistic sustainable development view is kept in mind in all design steps and stakeholders are actively involved.

This design often results in novel solutions that needs to be researched and developed before it can be launched into the market. This means that design and innovation are strongly connected. This connection is described in detail in Section 2.2. The sustainable design task, being much harder than conventional design, requires both creativity and integration of lots of information from very different knowledge domains.

Chapter 2 concentrates on the first step of setting up a (sustainable) design assignment. Figure 2.1 derived from Fig. 1.2 shows the set-up of a design assignment and the subsequent parts of Chapter 2.

After defining the challenges specific for a sustainable design (Section 2.1), and general guidelines on the design process as team work (Section 2.2), most of Chapter 2 is devoted to the problem definition step, which is very important in that the sustainable development vision is transferred into the goals, context and constrains. These subsections therefore

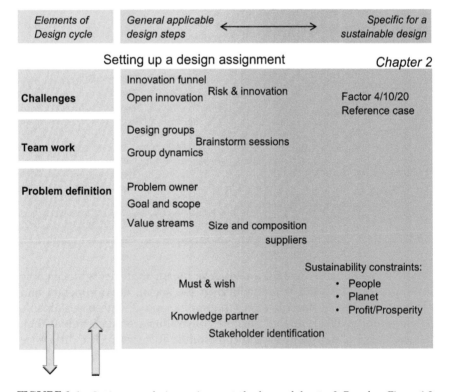

FIGURE 2.1   Setting up a design assignment: the focus of chapter 2. Based on Figure 1.2.

summarize the basic principles on establishing the problem definition, with a focus on the sustainability aspects. The principles are rubricated by setting the goal and scope of the design, including defining input and output value streams (Section 2.3), and setting constraints on sustainability (Section 2.4).

CHALLENGES FOR A SUSTAINABLE DESIGN

Factor 4/10/20 Challenges Relative to a Reference Case

At the start of the design, also called, the concept design stage, the design goal has to be challenging. This is needed to obtain a design that is significantly better than the existing process or product. This holds in general for any novel design. The reason for this is that if the design goal is not challenging, the design solutions will be only slightly different from the conventional design and at the end of the design effort, the conclusion would be that the benefit of the new design over the conventional technology is so small that the innovation effort cannot be afforded [33]. For a sustainable design, challenging targets can be set using the factors 4, 10 or 20 method [34], see Box 2.1. This can be obtained by taking the existing

---

BOX 2.1

IMPACT

A representation of radical improvement by the (qualitative oriented) IPAT or IMPACT equation {{52 Azapagic, Adisa. 2004}}, which reads $IM = P \cdot A \cdot C \cdot T$: if the impact on resources (Environmental impact, IM, equivalent/year) should stay constant within 20 years, a foreseen e.g. doubling in the number of people (P), together with a doubling in their average affluence (A, e.g. GDP/year) will require a fourfold improvement of the technology (T). The technology status is expressed as Environmental impact per unit GDP (E/$).

The difference between IPAT and IMPACT equation is the C of choice. This factor represents the choices society makes in types of technology to be applied. A country like Sweden has a low factor, as they chose technology with a low environmental impact. They choose for instance for well insulated housing with higher capital cost but lower energy cost. A country like the USA has a high factor. They chose low capital cost housing with little insulation, but with high energy cost for keeping the house at the desired temperature.

BOX 2.1 (*cont'd*)

The impact equation can be used to quickly estimate the environmental impact if the population or the affluence increases of a country. It can also be used to set targets for the technological innovation required to reduce the environmental impact increase caused by population and or affluence increase.

This equation is also the base for the factors 4, 10 and 20 improvement challenges.

The equation should be used with care. It suggests a linear relation between Environmental impact and the GDP. However, in reality, countries with the same GDP can have widely different environmental impacts. Luxemburg and the USA have comparable GDP/per capita however the hazardous waste per GDP of the USA is a factor 10 larger than Luxemburg [12], p. 115.

---

process or product as a reference case and stating specifically in what aspects the new design has to be significantly different.

A guiding principle for making a sustainable design is to choose a reference case from the existing processes or products and set challenging improvements for the new design compared with this reference case.

Stage-gate Innovation Funnel

A sustainable design is closely linked to innovation. This is best explained by describing first the stage-gate innovation funnel. This is a robust method to handle the design paradox [4,5,35—37].

Figure 2.2 schematically shows the decreasing freedom within the design process (red line), versus the invested time and money (black line). During the project time, the degree of freedom to make significant contributions to reduce investments and operational costs, and implementation of risk reduction, continuously becomes less, as project time proceeds. However, during the project time, not only the acquired knowledge, but also the time invested (and costs) increases. The operation window of innovation and also of making a sustainable design is in the initial stage (green shaded). Here, the effects of new ideas have the largest impact on the final design.

In the last two decades, innovation methods have been developed to partially break this paradox by two important measures. They are: 1.

Design & Development Paradox

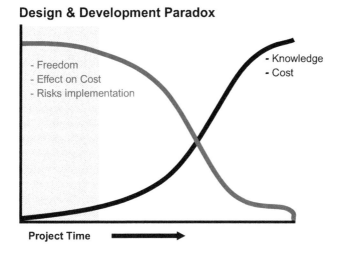

- Freedom
- Effect on Cost
- Risks implementation

- Knowledge
- Cost

Project Time

FIGURE 2.2 Design freedom, impact, knowledge and cost curves versus project time. The green shaded area is the most effective operation window for innovation paths and in making a sustainable design.

application of stages in the innovation path, and 2. subsequently putting the design central in each stage.

The breakdown of the innovation path from idea into stages, the first measure, uses gates in between the stages where a decision is taken to stop the project or to go to the next stage. The overall idea is that the most critical aspects of the innovation are tested first. If the innovation fails in these tests, then the project is stopped and little money is lost.

This stage-gate innovation funnel is very useful for the companies that want to contribute to the sustainable development, because they can place their sustainable development goals and criteria directly into the gate criteria. In this way all their innovations then become sustainable innovations, see also Box 2.2. For people working in a company with no sustainable business strategy still the proposed stage-gate criteria can be used, as these criteria are sensible in themselves and very useful to obtain a design result that is good for the company and society. So the individual innovator can communicate his method of working using this method.

There are many kinds of variations on stage-gate funnel methods [38–40]. These variations are streamlined into one sustainable innovation stage-gate funnel as shown in Fig. 2.3.

The sustainable criteria items are also indicated in Fig. 2.3 and will be explained in more detail in Section 2.3. Most important decisions for the

---

BOX 2.2

HINTS TO GENERATE INNOVATIONS THAT CONTRIBUTE TO SUSTAINABLE DEVELOPMENT

In the concept stage it is very useful to state high targets and very tight constraints in sustainable development terms. This forces the innovators to dream up radically new ideas, as conventional technology cannot meet these targets. Some companies even use sustainable development criteria, solely for the purpose of stimulating innovators to come to radically new concepts.

---

sustainable design solution are made in the first stage. Therefore the focus in making a sustainable design is on this first stage and that is also the reason that this book focuses on concept designing in this first stage and provides sustainable methods and tools for this stage.

Sustainable Innovation Funnel Stages and Gates

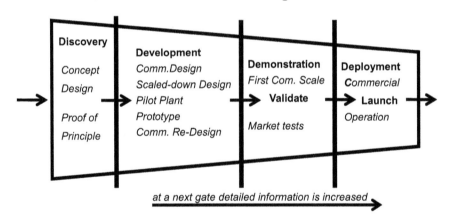

Gate criteria:

Fit Sustainable Need and Business Strategy?
Safe?
Healthy?
Environmental fit?
Social fit?
Rewards?
Risks okay?
Technically feasible?

FIGURE 2.3   Industrial innovation funnel with stages and gates.

This sustainable innovation method can then be used for all kinds of industrial branches as a starting point, see e.g. [41]. Individual companies may modify this method to suit their own particular need. A product innovation company for instance may want to insert an extra stage in the front end to focus first on linking local market needs with product functions.

The second measure to break the paradox of amount of information versus design freedom, is to give the design an important place in each stage of the stage-gate innovation funnel. The design serves two purposes. First, the design effort highlights the knowledge gaps which need to be closed by research and development and in this way only, the necessary aspects are researched. Second, the design is the means to communicate the state of the artifact at any point in the innovation path to both the team members and to the internal and external stakeholders. At the end of each stage, the design solution is used in the evaluation at the stage-gate. The stage-gate includes a meeting with the stakeholders and a decision is taken to enter into the next stage or not.

For sustainable innovation communicating the preliminary solutions with various stakeholders to ensure that the solution fits the social, economic and environmental context is even more important than the other types of innovation. Therefore this idea of using design as a communication vehicle is very useful.

In the first stage, the operation window in Fig. 2.3, all major Safety, Health and Environmental issues are to be raised followed by an assessment to resolve them directly in the first stage, or in the next stage. In general, it is better to resolve them in the first stage. Sometimes the issue can be resolved by changing the design. For example, a certain material used may have suspected eco-toxic properties. This requires additional research to determine the eco-toxicity and probably special additional design measures have to be taken to prevent even in small amounts of leakages to the environment. By making a design, taking extra efforts on safety or avoiding these materials removes the issue.

In the first stage also the technical feasibility of a process or product is to be determined to a certain extent by a proof of principle experiment. Only the most critical novel element of the concept is tested. In general this test requires a small experimental effort. The technical feasibility may also be tested by a simulation program.

At the beginning of the development stage a more detailed design is made, using results of (laboratory) experiments on specific subparts. All issues critical to satisfying the goals and constraints are addressed. At the end of the development stage, the detailed design is updated with all experimental and market research results. The designs in the development stage are sometimes called Final Concept Design.

In the demonstration stage an even more detailed design is made so that each part is sufficiently defined for purchasing and construction. For processes, it means that not only the major equipment is specified but also all the parts such as concrete floors, pipes, instrumentation, and housing are specified. For new products also the manufacturing process is defined in detail. Here, a clear communication with the engineering contractor is needed to ensure that in this stage also sustainable development criteria are applied.

Open Innovation with Sustainable Development Goals

Radical innovations contributing to sustainable development often means innovations across company fences. This means that in the innovation more than one company is involved. The radical innovation may also require competences found in institutes and universities. Innovations with intimate involvement of companies, institutes and universities are called open innovation. It differs from closed innovation where all work is done inside one company.

Figure 2.4 shows the roles and connections between the open innovation partners. In open innovation it is very important that all partners have the same goal and constraints in mind. For an innovation contributing to sustainable development, it is even more important that all partners share the vision of contributing to sustainable development and reach consensus about the main sustainable development targets. This

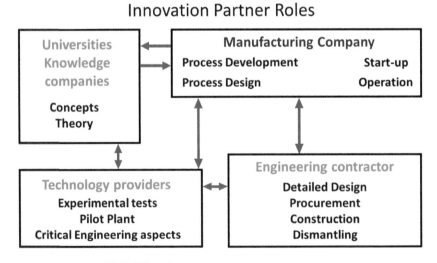

FIGURE 2.4   Partner roles in open innovation.

shared vision and consensus on the other hand will also strengthen the inter-company and institute collaboration.

The proposed generic sustainable innovation funnel and stage-gate criteria can be used as a communication tool to come to a joined assessment. At that stage in the innovation funnel a certain novel technology and a joined decision may be taken on the necessary steps toward commercial implementation.

A good book on the subject is [39].

Risks and Required Innovation Effort Constraints

Each new design concept will have a risk of failure when implemented commercially. This risk can be on any of the mentioned dimensions such as Safety, Health, Environment, Social acceptance, Economics or Technical feasibility. Table 4.2 in Chapter 4 shows how a rapid assessment on the risk of failure can be made. This risk of failure is reduced by going through the research and development stages, as indicated in the previous innovation section.

Risk has two underlying factors, the chance of occurring and the effect when occurring. Both factors can often be reduced. If the chance of occurring is very small but the effect is very large, such as with air borne nuclear radiation material escaping from a nuclear power plant, then it is better to reduce the effect by a large amount, rather than only reducing the chance of occurring. The reason for this is that very large effects such as nuclear air borne radiation affecting a whole country for decades can be seen as destroying the economic, the social and the environmental dimensions of a country, so are anti-sustainable.

In general, the higher the risk of failure the more R&D effort will be required to reduce the risk to acceptable levels. This relation can be used the other way around as follows. A company may want to set an upper limit to the total R&D effort in terms of money and time it wants to spend on the new design. Using Table 4.2 then the degree of novelty for the new design can be stated as a constraint. For instance, a new product should be designed but its manufacturing process should not contain novel elements, but only the conventionally proven elements.

In this way the research and development effort in terms of cost and time is determined up front of the actual expenditure. The relation between degree of novelty and R&D effort is not a very accurate and reliable relation. Therefore along the R&D project in time the actual R&D cost should be compared with the predicted cost and also the R&D cost forecast should be regularly reviewed. If the predicted cost is higher than the stated limit and if the anticipated profit is not increased, then the innovation will stop at that particular stage-gate.

In summary, a challenging goal and very tight constraints compared to the existing solution will result in a very novel design, which will require a much larger innovation effort than a moderately defined goal and constraints. The estimated required effort should be compared with the effort the company is prepared to accept and to the expected profit that can be made.

DESIGN PROCESS AS TEAM WORK

Forming Design Groups

The size of a design group is the first decision to take. Usually design groups have sizes ranging 2–8 for which groups of 4 participants in general perform the best. Smaller groups lack diversity of thinking and behavior. Larger groups of 6 participants perform reasonably well, but often one or two participants travel along rather than being critical on intermediate results and having less input.

Working in complementary groups with a variety in background, culture and gender create far better design results and the participants learn far more on design. The reason is that they have to listen more to each other, because different views are expressed, which can only be understood by asking lots of questions. This questioning and answering improves the idea formation and group coherence. So designing the groups rather than letting the participants choose their own group members is a good practice.

Also a fruitful method of designing successful groups is using Belbin team roles, shown in Box 2.3.

The Belbin roles are categories of human behavior in teams. It means that most people have a natural tendency to play one of these roles in a team. When asked most people, they indicated quickly which role describes their team behavior best. Belbin found out that teams perform best when all roles are present in the team and teams perform poorly when only one or two roles are present in the team.

For small teams the Belbin roles are clustered as shown in Box 2.3 into 3 main roles: Chairman, Input-Output creator and Controller. The team should have these 3 roles. This means that the chairman of the team should be selected if his natural role is one the roles described in the first cluster. The same holds for the other clusters.

For groups between 4 and 8 people the other roles may be distributed according to the type of problem to be solved. The role of the specialist in the team should be considered with care: this role can be a strength but also a weakness, as the specialist tends to focus his specialist knowledge only and even prioritizes this over the team progress. This role is not

---

BOX 2.3

THE BELBIN TEAM ROLES AND THEIR CLUSTERING

Cluster: Chairman
   Coordinator ensures focus on goals and task distribution
   Implementer provides a practical strategy and carries it out.
   Team worker gets members to a team
   Shapers drive team to result
Cluster: Input to Output creator
   Resource investigator gathers facts and ensures that results are brought to outside world
   Plant is a creative unconventional problem solver
Cluster: Controller
   Completer-finisher polishes and scrutinizes the work for errors at the end of the job.
   Monitor evaluator makes impartial judgments
Advise to call in specialists as advisors and not a part of permanent team
   Specialist provides in depth knowledge in a key area.

Source www.belbin.com/rte.asp

---

essential in the team. Specialists should be consulted at specific moments to provide key information on certain aspects of the design.

Group Dynamics

Designing is mainly team work, so it is recommended to pay attention to the well-known observation that the teams go in general through the following phases:

Forming: Formal group formation can be best performed by others, in an industrial setting it may be the appointed team leader in co-operation with the internal client, also called problem owner. In a course setting, this can be done by course leaders. Often the team members are selected on their specific technical skills. However, it is also important that personal characters are complementary according to the Belbin roles. Therefore the team member selection should be also based on complementary Belbin roles to ensure good team work.

Forming also involves tasks distribution. This is in general done by the team itself. A guideline is to appoint a chairman for the meetings. This chairmanship may change per meeting. The chairman ensures that each meeting has a specific goal and a clear agenda with a timeline and that the meeting sticks to these.

Also an important task for meetings is to appoint a secretary for recording decisions, actions and plans.

Further tasks may be defined and distributed over the team. However, it is important that all the members have the same overall goal in mind. It is therefore important to have team meetings to integrate part solutions into the overall concept.

Storming: Storming is the phase of intense discussions between team members and is sometimes chaotic and sometimes resulting in angry opponents shouting at each other. A brainstorm procedure as described in Box 2.4 will be of help to facilitate the idea generation. In reality this phase is needed to learn to know each other, to obtain a position in the team and to generate rules of engagement. It may last for a long time or

---

BOX 2.4

BRAINSTORM RULES AND GUIDELINES

1. Appoint a brainstorm facilitator who guards the brainstorm process
2. Do not criticize any idea
3. State the problem to solve as a goal, but do not mention many constraints
4. Generate many ideas, especially strange ridiculous ideas
5. Share all ideas with all brainstorm members (by putting all ideas on flip-overs)
6. Enrich and expand ideas
7. Cluster ideas and enrich again
8. When the flow of ideas stops make a break
9. Insert a session where only wrong, not workable ideas, are generated
10. Insert a session where the complete opposites of the wrong ideas are generated
11. Make sure that all ideas are collected and stored
12. The designer outside the brainstorm session incorporates ideas into the design
13. Give feedback to the brainstorm members what has happened to the results

Hint: Make sure that the brainstorm group does not contain supervisors of any of the brainstorm members. These supervisors often unconsciously prevent the flow of weird ideas.

BOX 2.5

INCUBATION TIME AND PARKING LOT FOR IDEAS

When focusing on defining the design problem often already solution ideas pop up in the designers mind. This may be during the problem definition activity or often later, outside the working hours. This time is called the incubation time. Subconsciously the mind keeps working on the problem [90], p. 149. These ideas should be directly stored and be parked during the problem definition stage in a document section called: Parking lot for ideas.

---

for a short spell. Knowing the occurrence of this phase may help to ease the tension.

Norming: This is the stage in which group norms are established by which the team will operate for the remainder of the project. Often it helps to write down these norms. They may include: we want to learn from each other, so we listen and if we do not understand what someone is saying, then we ask for clarification.

Performing: The phase in which the team acts together, using the strengths of each member to generate the best result possible.

Lecturing the participants up front of their team work about these team dynamics helps in general to go faster through the first phases. Finally, Box 2.5 provides some practical hints for individually storing ideas.

SETTING GOAL AND SCOPE

Assigning the Design Problem

Initially, the problem definition will be far from complete. The main reason is that it is hard to think in all aspects, because this requires a lot of knowledge from very different areas. Even if the problem is defined by a problem owner in an industrial setting, the initial problem definition will not be complete. To emphasize this, often this first problem definition is called the primitive problem definition. The designer or design team should be aware of this fact and should revisit the problem definition every time progress is made in the design and discus the improved problem definition with the problem owners and other relevant stakeholders. In this way, the problem assignment is improved during the design. For a sustainable design, problem definition will be even more important, as more aspects of the context in which the design solution has to fit have to be taken into account. The designer can highlight the importance of obtaining a sustainable design solution and that the sustainable development focus should start with the problem definition.

Goal of the Design

The goal of the new design should be clear to the designer. In general, the goal of the design is a simple statement (and all other requirements are stated as constraints). A process example is: We want a new sustainable process for product P for the capacity of 100 kton/year. A product goal example is: We want a new airplane for 300 passengers that has a factor 2 lower energy requirement per kilometer than the best conventional airplane of the same size.

For this goal setting the market need for a design has to be firmly established, otherwise in later stages both the design team and the internal customer will get disappointed or even worse, the final constructed solution appears not be needed in the market. There are some differences between a product and a process focused design: for a product, a market survey of some kind is required. For a new process design its real need may be obtained by asking the wishes of the client. Sometimes a better defined need is only established if the first design solution (based on the primitive problem definition) is presented to the client and the other stakeholder in the company and then they are often able to state their wishes.

For designing a sustainable government policy, goal and scope setting is enormously important. Successful policies have in general a people behavior change aspect and an innovation aspect. The so-called IMPACT expression shown in Box 2.1 can be of help in goal setting for behavior change (Choice) and for technology innovation (T). If for example the total net carbon dioxide emission to the atmosphere by a country for person transport has to be reduced by a factor 2, say 20 years from now and the population growth is known and for simplicity, take it here as zero, then the policy goal can be quantitatively defined as 50% net carbon dioxide reduction. The scope of the policy problem definition can then be that, both the behavior chances and technology innovation on person transport will be a part of the policy. A hearing with politicians of all parties may be helpful to set goals and scope. Consensus on the goals and scope are hard to obtain in this stage of policy making, but consensus on both the technical innovation and behavior changes pursued may be achievable.

Design Scope

For a design it is also important to determine the scope of the design. Table 2.1 shows 6 levels in the process and product industry,

TABLE 2.1   System Levels in the Industry and Options to Improve

System levels	Options
Life cycle chain	New feedstocks and products
Industrial park	Waste is feed integration
Process	New design or revamp
Unit operations	New design or revamp
Equipment	Select
Operation	Quality, control, motivation

which should help in defining the design scope. It will be clear that a wide design scope such as including several steps in the life cycle or designing a new industrial park has more potential for obtaining large improvements in the direction of the sustainable development than a small design scope such as a new unit operation of a single part of a product. At the level of industrial park waste streams of process A can be used as feedstock for process B. Heat generated in process C can be used in several other processes. Examples can be found in [141].

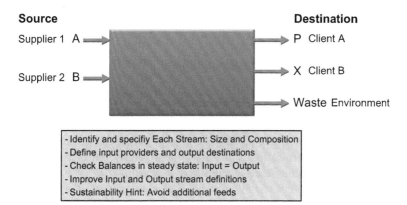

Define Process System Boundaries Design Object by all Input and Output streams

Source	Destination
Supplier 1 A →	→ P Client A
Supplier 2 B →	→ X Client B
	→ Waste Environment

- Identify and specifiy Each Stream: Size and Composition
- Define input providers and output destinations
- Check Balances in steady state: Input = Output
- Improve Input and Output stream definitions
- Sustainability Hint: Avoid additional feeds

FIGURE 2.5   Input and output streams and supplier and destination determination.

Defining Value Streams

Subsequently, all input and output streams should be defined on needed size and composition. Fig. 2.5 shows how all input and output streams can be represented with their sources and destinations. Defining all output streams and all input streams in composition and size is an essential element of the problem definition. In general, it is most convenient to start with defining the main product output stream. Then main input streams required for that product output stream can be defined. Finally, all other output streams and additional input streams are defined.

Here sustainable development items related to environmental emissions over the life cycle (see chapter 4 on life cycle assessment method) can be made effective by allowing only output streams that have value and not allowing waste streams. In this way the design solution has to ensure that no waste streams occur. This is in line with the concept Cradle to Cradle [7]; explore options to create only output streams of value to others.

To ensure that the output streams have indeed value specific destinations (customers) for these output streams have to be found. Figure 2.5 shows a presentation of the input and output streams with sources and destinations. In general, waste streams should be avoided, as feed streams from depleting resources. For the selection of feed streams and handling waste streams, the sustainable development comes directly in view. Again the whole life cycle should be taken into account for selecting sustainable feed streams. Chapter 4 shows a rapid Life Cycle Assessment suitable for concept design purposes.

What also needs to be defined are the suppliers from where the inputs are to be obtained, or a list of suppliers with the feed specifications.

SUSTAINABILITY CONSTRAINTS

Now that the goal and scope is defined, all constraints can be set. In sustainable concept design the following guiding principle will be of use.

Guiding Principle for Sustainable Design

Leading principle: think of a new sustainable context, a new vision about the future in which the design will flourish. Think holistically about future very stringent laws on health, safety, and environment. Agree with the customer on this vision and on the resulting quantitative constraints for the design. Always express environmental emission constraints in terms of the cradle-to-cradle Life Cycle. Read chapter 3 and envisage for each context level how the new design should be shaped.

In particular for companies that want to differentiate themselves from others by having sustainable development in their strategy, this leading principle is very important. It can be used to design new business models, new product design classes and new types of services. Think for example of the green car service, where personal transportation by car is decoupled from car ownership. A new way of laundry washing where both the washing powder and the washing machine are integrally optimized.

Experienced designers will have generic constraints list to communicate with the clients to obtain more specific and quantified constraints list. The items below can be used in the same way. They are meant for the engineer who wants to design with the sustainable development goals and criteria in mind.

In design often hard **must** constraints and nice to have, **wish** constraints are distinguished from each other. In the first round a design is generated that meets only all those hard must constraints. In the later rounds the design is optimized using the wish goals and constraints for which tradeoffs are made. The so-called Pareto optimization may be followed in which the design parameters are varied to maximize on all wishes. The Pareto optimum is reached when by further variation some wish output parameters improve, while others suffer. However, in radically novel sustainable designs often the improvements are made on all constraints by one integral novel concept as shown later in this chapter in the section synthesis. This type of solution can be stated in the problem definition, as the new design must be far better on all constraints.

People, Planet, Profit/Prosperity

The Triple P, People, Planet, and Profit or Prosperity requirements are providing a well working metaphor for the set of social, ecological and economic constraints for a sustainable design, see discussion Section 1.2.

Criteria for human safety and human health are strongly related to social acceptance; the *People* dimension of sustainable development. Stringent constraints for safety will increase the chances of acceptance by society and is therefore a very important aspect of a sustainable design. Constraints in the concept design stage related to this subject are mainly about human toxicity and explosion hazards of the chemicals for feedstock, in the process and in the products. A must criterion can be that the toxicity of all components applied in the product and process design should be known, and their toxicity should be much lower than for hazardous components of the existing process and product. Toxicity data may be quickly found using REACH or IENEX websites (see Box 2.6). For food products or products that are used in the food industry like package material food approval by the Food Approval Authority (FAO)

BOX 2.6

REACH AND EINECS FOR SAFETY DATA

REACH is the European Community Regulation on chemicals and their safe use. Its acronym stands for Registration, Evaluation, Authorization and Restriction of Chemical substances. The REACH law entered into force on 1 June 2007. A central database run by the European Chemicals Agency (ECHA) in Helsinki can be consulted for safety and health information on many chemical substances. Alternatively the EINECS (European Inventory of Existing Commercial chemical Substances) list can be consulted. Avoiding the use of a very hazardous chemical like phosgene, can be a design criterion for a new process for a chemical company to keep a license to operate. A comprehensive book on this subject is [127]

---

is required. A list of approved substances is provided by websites of the FAO.

The *Planet* gives constraints about emissions to the environment, which should always be defined in relation to the whole Life Cycle from cradle-to-end of life cycle and back to cradle. Chapter 4 describes a rapid life cycle method and shows environmental emission types.

Often the mistake is made that the constraint is defined only in terms of the life cycle of the design object. This in general leads to the wrong solution in terms of sustainable development. Take for instance the design of a novel aircraft in which only the emissions to the environment from resources to the aircraft construction are taken into account and that the constraint is stated as the emissions to the environment should be factor 2 lower than the existing aircraft. Then the aircraft may be designed using a construction material with that factor 2 lower emission to the environment. If however the total life cycle is taken into account, including the use of the aircraft for 30 years then for instance the carbon dioxide emissions during those 30 years will be the overriding factor and if flying the novel aircraft is less fuel efficient then the emissions to the environment by this aircraft increases, instead of decreasing.

Also a trap is to set a few constraints very tight, while other aspects of the design are not stated as constraints. For instance a very tight constraint may be on global warming gas emissions while a constraint on aquatic toxicity is forgotten.

Finally, economic constraints (*Profit/Prosperity*) can be defined in various ways. The easiest constraints to apply are about lower cost than the reference case. For processes this is an easy to use constraint. For products it is only a useful method if the product is directly comparable with a similar product in the market. For novel products the market price and the profit per product may be set by the marketing department. The design has then a target cost price which is set as a hard constraint.

Required Domain Knowledge Partner and Stakeholder Identification

A part of the problem definition is also identifying what type of knowledge; also called domain knowledge will be needed to solve the design problem. Particularly in radically novel designs with complex interactions between design elements on the resulting performance it is difficult to identify what type of knowledge has to be obtained to be able to design. The following simple step method works very fruitfully in identifying what knowledge needs to be obtained.

- Step 1: Study all goals and constraints, as defined in the problem definition.
- Step 2: Guess what constraints will be most difficult to meet.
- Step 3: Determine which knowledge domains will help in meeting those constraints.
- Step 4: Obtain knowledge from literature and potential partners, such as industrial companies and university groups.

The human mind is capable of subconsciously thinking through enormously complex problems. This subconscious thinking works best, by first focusing strongly on Step 1 so that some anxiety occurs. Step 2 is then often easily taken. Even inexperienced students quickly feel what constraints are hard to meet. Step 3 is then in general easily taken and a short list of required knowledge domains is quickly made. Step 4 is then also easy as now keywords for literature searches and potential partner identification are available.

The cradle-to-cradle design method can also be of use (see also chapter 4) to generate a list of acceptable materials to choose from [7].

Now that the problem is defined in terms of goals, scope, input and output, context and constraints identifying all relevant stakeholders for the design problem will be easy. Chapter 3 will also be helpful in imagining the design context and the relevant stakeholders. As a check for each type of constraint a stakeholder should have been identified.

Structuring the Sustainability Context

An important part of the sustainable design problem definition is identifying all relevant context aspects. To facilitate this identification, we structured the sustainable design context in four context levels:

- Planet
- Society
- Business
- Engineer

Each context level is separately discussed in sections of this chapter.

Sustainability starts on the level of *Planet*, with resource depletion, equity, and biodiversity, related to an expected increase in human population and welfare. Then, on a more local level, *Society* has a practical application of sustainability, including government, NGOs, and engineering associations. As a separate level, *Business* has an increasingly and in some branches even leading role in sustainability.

A sustainable design starts at the level of the *Engineer*, usually working in teams. The first three context levels can be regarded as boundary conditions, varying from guidelines to strict requirements for the design engineer. However, at the level of the engineer, personal convictions and the dynamics of teams also determine the decision processes.

BACKGROUND OF CONTEXT LEVELS

As was pointed out in chapter 1, sustainability requires a thorough look at the system in which the engineer is operating. Figure 3.1 shows the subpart of Fig. 1.2 of chapter 1. The General Living System of J.G. Miller [19] [16], p. 418, as a social system defined on observations from nature, functions as a leading principle. The lowest level of Miller's levels of generalized living systems, the cell and organ are not included. The two levels above, of the individual and group, are combined in the level

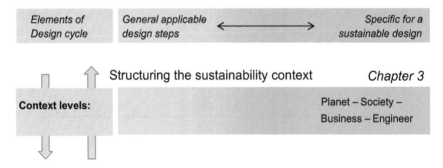

FIGURE 3.1   Subpart of Figure 1.2, denoted the contents of chapter 3: Structuring context on sustainability in four context levels: Planet, Society, Business, and Engineer.

Engineer. Miller then defines the organization, which here is directed to *Business.* Finally, the two highest levels in Miller's system are in this book defined by *Society* and the supranational level is defined by *Planet.*

These four levels describe the relation of the engineer with the context of making a sustainable design. The difference between a sustainable design and a usual design is that sustainability adds more and other requirements from the mentioned context levels. Therefore, the definition of a sustainable design (see chapter 1) starts with *Without compromising usual design criteria such as costs, appearance and quality*: engineers as designers always have to deal with the classical view of serving clients, which are the society at large, their clients or employers, and their profession [42,43]. The section on the fourth context level, *Engineer,* elaborates on the position of the engineer with respect to duty to incorporate sustainability in a design. The other three context levels describe the context of a sustainable design, reflected in the definition of a sustainable design as: *all environmental impacts of a design outcome throughout the complete life-cycle should be taken into consideration, including social and economic well-being, such as ethics and the environment.*

CONTEXT LEVEL PLANET

With the context level Planet, not only the geometric scale of the planet is meant, but also the temporal scale; working on sustainability is for the next generations in future times. This intertemporal [44], p. 344, or intergenerational [15] p. 61, equity is part of the precaution principle. Humanity has to take considerations of living species and nature in general in their ethical assessment, and not only of issues directly concerned to humans [15], p. 49, [42]. Ecology and environment came into focus since Silent Spring of Rachel Carson [45] p. 4, and Limits to Growth [46]. Although daily practice developments on a global scale usually may not directly affect decisions concerning design assignments, the cognizance of signaling the impact of the design on the Planet is important nowadays, which includes awareness of global trends [47], also see Section 3.4. Sustainability is seen as being typical for the 21st century everywhere. Serious and well-documented literature describes fact by fact the unstable and unsustainable current global situation. Overviews and finding new ways are available from environmental science [45,48–50], leading economists [21,51–54], and from leading experts [3,17,34,55,56]. In this respect, the reports of the Worldwatch Institute, summarized in their State of the World, are milestones. Since 1984, these books provide overviews, annually grouped by topic, on *Innovations that Nourish the Planet,* [57], *Transforming Cultures from consumerism to sustainability* [58], and *Into a Warming World* [59]. Not only the current problematic state of these subjects is rubricated, but experts also explore possibilities to find a way out. These and other observations undeniably show the Earth's limitations: the work of an engineer in daily practice, in developing new technology, will be influenced by these trends. The factors concerning the anthropocentric influence on the Planet are divided in two sections: the safe operating space for humanity and more specific, on the global carbon cycle and energy scenarios.

Safe Operating Space for Humanity

The effect of mankind on the global environment was negligible in the period of the Holocene, starting 10 000 years ago until the Industrial Revolution in the eighteenth century. However, since the Industrial Revolution, human actions have become the main driver of global environmental change. Many subsystems of the Earth are behaving non-linearly: If thresholds are crossed, the subsystem may become unstable and could shift to another state, often potentially threatening human living conditions [60]. The main subsystems are summarized in Fig. 3.2 [60]: the

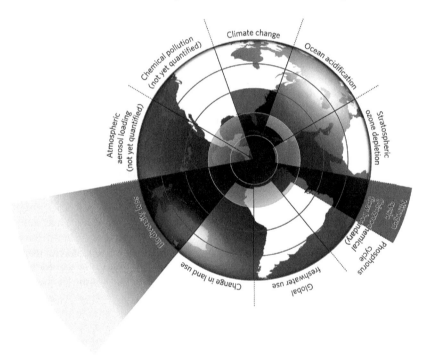

FIGURE 3.2   Safe operating space for humanity. The inner green shading represents the proposed safe operating space for nine planetary systems. The red wedges represent an estimate of the current position of each variable. The boundaries in three systems have already been exceeded, see the main text. Reprinted by permission from Macmillan Publishers Ltd: *Nature*, copyright 2009 [60].

categories are quantified and schematically show the level for which the Earth is expected to recover from, called the safe operating space.

Three categories seriously exceed the green inner circle, which are biodiversity loss, the biogeochemical flow boundary of nitrogen, and climate change. These three are discussed hereafter.

Biodiversity loss has increased enormously during the last few centuries. At this moment, about 25–30% of all known species of mammals are at risk of extinction, as well as 11% of the known birds, 20% of the reptiles, 25% of amphibians, and 34% of fish, primarily freshwater fish [45]. Apart from the usual irreversible aspect, it is known that a rich mix of species underpins the resilience of ecosystems. [60]. The studies done on behalf of the large scientific project Millennium Ecosystem Assessment reveal all trends and threats regarding the ecosystem [61].

The global **nitrogen cycle**, connected to the food cycle, shows how success and its negative consequences are closely interlinked on a global scale. Food production depends on the cultivated area, irrigation, and nitrogen input as artificial fertilizer. The food production has tremendously increased in the last few decades, see Fig. 3.3 [62], which shows the trends in the factor related to global food production from 1961 to 2003. The effect

Sources: FAOSTATS, SOFI, Millennium Ecosystem Assessment

FIGURE 3.3   Global trends in food production [61].

of an increased world population is reflected in the food production per capita, which only slowly increases despite the threefold increase in food production. The Green Revolution managed to double the grain yield from 1960 to 2000, but the new crops demand large quantities of water (and fertilizer). However, the total input in nitrogen was at the beginning of the 21st century comparable to the total amount of nitrogen already present in all natural pathways, and this is expected to grow further, see Fig. 3.4 [62]. These trends on food production have consequences on a global scale on nature, in particular on loss of habitat for endangered animal species. Food production is present in many aspects of the society, and therefore process and product design in this area are important. Books of more specific knowledge, e.g. when working in the food industry, such as Micheal Pollan

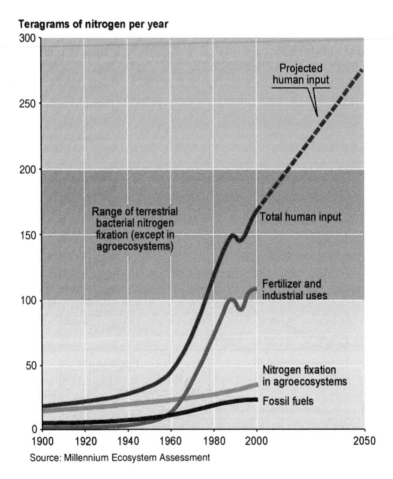

FIGURE 3.4 Nitrogen introduced by agriculture compared to nitrogen already present in the biological cycle [61].

[63], provide very interesting and fundamental information on the current trends in global food production.

Food production is also connected to energy usage. Food production consumes much energy in the life cycle chain in the production of fertilizers and in transport. With coal and oil as the main fossil-fuel-based energy suppliers, the energy consumption has steadily increased during the past few decades.

Climate change, closely coupled to enhanced global warming, is recognized as a serious human influence on the Earth's systems. The main source of enhanced global warming is burning fossil fuels and thereby emitting carbon dioxide. This causes an increase of the greenhouse gas carbon dioxide concentration in atmosphere. In absolute terms, the amount of carbon present in the atmosphere mainly as carbon dioxide is 0.039% (390 ppm) of all gas molecules. It has increased from 280 ppm before the Industrial Revolution to the present level and is still increasing. It is the main contributor to enhanced global warming. An impression of how much the carbon dioxide in the atmosphere can increase by fossil carbon burning is obtained by comparing the present total amount of carbon (as carbon dioxide in the atmosphere of about 500 Gton [64]) with the amount of carbon in the Earth's crust. The deposits of carbon in the Earth's crust for commercial exploitation is around 100−200 Gton for oil and natural gas and 5000 Gton for coal. So, the latter alone is able to accomplish a tenfold increase of the carbon content in the atmosphere. Methane and nitrogen oxide play an important role as well in enhanced global warming, because their greenhouse gas potential is far higher than carbon dioxide [49], p. 125 [45], p. 482.

Global Carbon Cycle and Energy Scenarios

The carbon footprint is accumulating all global warming gas emissions by mankind's actions and express them as carbon dioxide equivalent mass units emissions [65], see text Box 3.1. There is massive evidence that

---

BOX 3.1

CARBON FOOTPRINT

The six greenhouse gases of the Kyoto protocol all add to the carbon footprint, which measures the total greenhouse gases of a product, event, or organization. Apart from carbon dioxide, the gases are methane, nitrous oxide, hydrofluorocarbons, perfluorocarbons, and sulfur hexafluoride. The greenhouse potential of the gases is recalculated to tonnes of carbon dioxide.

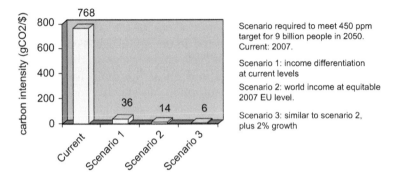

FIGURE 3.5   Required carbon intensity in 2050 to meet 450 ppm target of atmospheric CO_2 concentration at different scenarios [142].

through burning of fossil fuels by mankind enhanced global warming will affect the climate on the entire planet. From this climate change, the poor will suffer most and first [56]. Climate models indicate particular vulnerability in the tropics, where the food supplies already is or will be undermined for hundreds of millions of people [59], p. 8.

Figure 3.5 shows the challenge with respect to climate change. The 2007 carbon intensity in gram carbon dioxide per unit of currency ($) is shown for the world scale, UK, and Japan. Although the carbon intensity has reduced by around 40% from 1980 to 2007, there is a worldwide trend of an increasing usage of fossil fuels and subsequently CO_2 emissions, which appears to follow the increasing world GDP, see Figure 3.6. The four bars at the right end of the graph are taking into account these trends, in giving

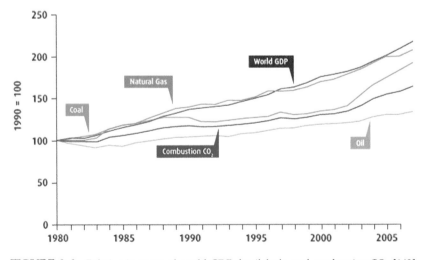

FIGURE 3.6   Relative increase of world GDP, fossil fuels, and combustion CO_2 [142].

indications for different scenarios for the required levels to stabilize in 2050 at a level of 450ppm carbon dioxide equivalents. The latter is looked upon to have a relatively mild effect regarding climate change. Scenario 1 is quite optimistic in assuming a moderate growth of 0.7% in population and 1.4% income growth, which results in an allowed carbon intensity of less than 40 $gCO_2/\$$ to stay below the 450 ppm level. However, if the population grows to 11 billion people (scenario 2), the carbon intensity should be less (below 10 $gCO_2/\$$). Assuming that in 2050 the income between developed and developing world is comparable (scenario 3), also with a 2% worldwide growth (scenario 4), the calculated carbon intensity per \$ should be decreased by a factor typically in the range of 50–100 (see Fig 3.5, from above 750 to below 10 $gCO_2/\$$). The conclusion of this exercise is that, although the exact numbers may be under discussion, the challenge of reducing the carbon intensity is anyway enormous.

Thereby fossil fuel is technically a very attractive energy carrier due to its high energy content per mass unit. While oil in particular is attractive because it can be pumped, a replacement with other fuels is not easy, which is also depicted in an energy scenario of IEA (Fig. 3.7[66]) showing that fossil fuels are likely to stay the main source of energy for a long time. The 450 PS is denoting the policy scenario for not exceeding 450ppm CO_2 level, which requires tremendous investments [67].

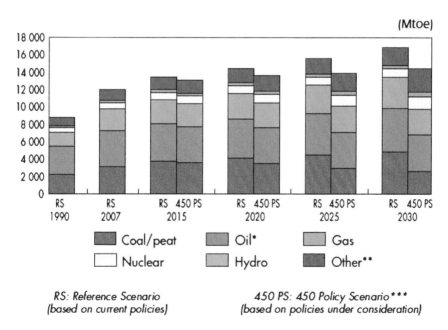

FIGURE 3.7 Primary energy by source [66]. *Key World Energy Statistics* © OECD/ International Energy Agency 2010, page 46.

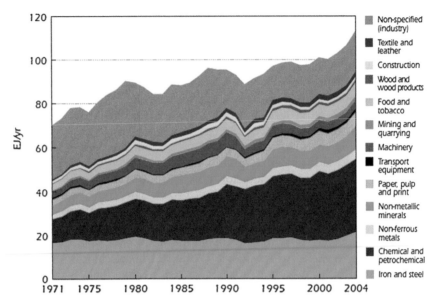

FIGURE 3.8   Energy usage of industrial sectors [68]. *Tracking Industrial Energy Efficiency and CO$_2$ Emissions* © OECD/International Energy Agency 2007, page 41.

Figure 3.8 shows that the energy usages of various industrial sectors increase as a function of time. It is therefore clear that the global energy usage and the carbon cycle are a major issue on the scale of the planet [68].

Summarizing the context level the Planet: most important threats for the planet and its living species are loss of biodiversity, the extensive amount of nitrogen added to the biological systems, and the enhanced global warming, due to carbon dioxide emissions by burning fossil fuel, all due to large-scale activities of mankind.

CONTEXT LEVEL SOCIETY

In this section, two subjects are discussed, which are closely connected to engineers' work in relation with society. First, the main basic human needs and the United Nations millennium goals are discussed in relation to what engineers have to do. Second, the 'Grand Challenges for Engineers for the 21st century', as formulated by the National (US) Academy for Engineering and the common parts are discussed.

Basic Human Needs

The Human Development Index (HDI) of the United Nations exhibits four key aspects on the basics of human quality of living: infant mortality, population growth rate, longevity, and (il)literacy[49], p202. Every world citizen can be believed to strive to an HDI of close to one, implying a fulfilling of the basic needs. Figure 3.9, showing both the HDI and the ecological footprint, reveals the desired area to be in [69,70].

As can be expected, the four key aspects also show a strong correlation with the gross national product (GNP), [21], however, to a certain point. Figure 3.10 gives another indication of welfare, in terms of subjective well-being. After a certain point, the increase in the percentage happy and satisfied with life does not substantially increase anymore [71]. As GNP is closely coupled to the carbon intensity, more goods, energy usage, and therefore increased carbon intensity do not imply to add a broader feeling of well-being.

United Nation Millennium Goals

As the leading organization of nations, the United Nations (UN) focuses on the large differences between populations. For 80% of the world, the 'needs of the present' are quite basic, and has nothing to do with the welfare of the Western world, especially for the 1 billion undernourished. The UN committee has stressed the importance of the common future of our world, for every inhabitant.

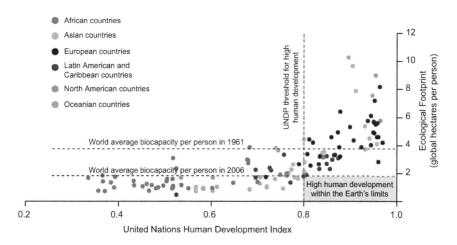

FIGURE 3.9 Meeting the dual goals of sustainability: high human development and low ecological impact, depicted for a snapshot of countries as the Global Footprint Network's Ecological footprint at the United Nations Human Development Index (HDI). An HDI above 0.8 is defined by the United Nations Development Programme's as an high human development. In the box in the right corner an HDI above 0.8 can be achieved within the Earth's limits. © Global Footprint Network (2009). Data from Global Footprint Network National Footprint Accounts, 2009 Edition; UNDP Human Development Report, 2009 [69].

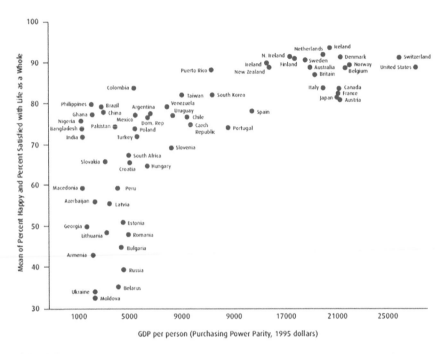

FIGURE 3.10   Happy and satisfied with life as a whole at different levels of income. Source: Worldwatch Institute, *State of the World* 2008 [71].

Related to human well-being, the UN has defined eight Millennium Goals, monitored by the World Bank [72,73]. Seven of the eight Millennium Goals are closely connected with poverty, see Box 3.2. Only Goal 7 is directly related to environment, the other are mainly focused on human

---

BOX 3.2
─────────

UNITED NATIONS MILLENNIUM GOALS

1. Eradicate extreme poverty and hunger
2. Achieve universal primary education
3. Promote gender equality and empower women
4. Reduce child mortality
5. Improve maternal health
6. Combat HIV/AIDS, malaria, and other diseases
7. Ensure environmental sustainability
8. Develop a global partnership for development

development, usually summarized with the large population living on 2 dollar or less a day implying a living below all standards. This care of world's poor is also recognized by the Brundtland committee (see box 1.2). The need of water is depicted in the seventh millennium goal: the UN uses the word sustainability in terms of the availability of freshwater, also needed for irrigation. In the clarification of the Nobel Peace Prize 2007 for IPCC and Al Gore, the Nobel Prize committee mentions the Sahel region as an example where conflicts on water resources have led to wars, which gives the connection to the Nobel Prize for Peace. On a world scale, UN defines that main problems and challenges are connected with inequity and undernourishment or food production.

Grand Challenges of National Academy of Engineering

It is interesting to compare the Millennium Goals (MGs) of the UN with the 'Grand Challenges' of the National Academy of Engineering [74,75]. The 'Grand Challenges' (see Box 3.3) imply the need of urgency of making intelligent designs which help to solve complex problems. The list of challenges show the complexity of present-day problems, with great economic and universal social impact [76,77]. A couple of the challenges are related to the global environmental problems mentioned in Section 3.1, such as to manage the nitrogen cycle, provide access to clean water, and

---

BOX 3.3

GRAND CHALLENGES FOR ENGINEERS

Make solar energy economical
Provide energy from fusion
Develop carbon sequestration methods
Manage the nitrogen cycle
Provide access to clean water
Restore and improve urban infrastructure
Advance health informatics
Engineer better medicines
Reverse-engineer the brain
Prevent nuclear terror
Secure cyberspace
Enhance virtual reality
Advance personalized learning
Engineer the tools of scientific discovery

Source: [74]

advance personalized learning, while others seem to be more related to the problems of the developed countries, such as to restore and improve urban infrastructure, prevent nuclear terror, and secure cyberspace.

From this list, challenges as to provide energy from fusion, which for many decades already is a promise for the future, may be viewed upon as a limited view regarding sustainability. One can wonder if the (huge) money involved cannot be better spent on more urgent matters.

Role of Government

Within society, the government is obviously also an important player on sustainability. The government always has had the task of balancing the individual freedom against the social good [21]. Individuals generally strive at a personal well-being and welfare, showing not much consideration with common goods [21], and also, the *Homo Economicus*, as it is by economists often referred to, behaves as a person in community [52], p. 164, rather than a solitary operating individual. Therefore, in terms of sustainability, modern government policy acts as a 'choice editor' [58], in supporting sustainable products or processes, by legislation (e.g. green labeling of houses, as LEED in the USA and in parts of Europe) or by taxing [58]. Another view on the government is their role as a moderator. There are many cases available in which the (local) government acts as a moderator between (market) parties as a co-designer of sustainable development [50] and as a designer of legal structures to repair errors. As a starting point, some society errors are shown in Box 3.4.

Nongovernmental Organizations (NGOs)

Nongovernmental organizations also play a role in shaping a sustainable society. They (claim to) speak and act on behalf of parties with no voice, such as the environment, or the poor. Examples of global NGOs are Green Peace, protecting nature and Oxfam, fighting poverty. NGOs have in general a single-issue viewpoint. They try to influence governments, businesses, and civilians by all kinds of means. Because of their single-issue viewpoint, they can clarify certain aspects of a problem, and also because of their single-issue viewpoint they cannot strike a balance between all relevant stakeholder parties. This has still to be done by government.

CONTEXT LEVEL BUSINESS

Sustainability Challenges

There is a need for transformation of business in relation to society [2,58]. The role of business is important as companies direct large groups

---

BOX 3.4

PRESENT ERRORS IN SOCIETY SYSTEMS

- The current waste of energy in terms of fossil fuel is enormous as the conversion of primary fuel resources to useful energy sources on the average is only around 40–60% [128].
- Less well known, but very prominent, is that around 80% of the materials connected to the end (consumer) product is already consumed by vendors in the supply chain [4,65].
- Connected to the material and energy flow is the (almost) free usage of common goods, as air and water. Economic models and key figures generally focus on revenues of products, in which waste is not incorporated. In this respect, waste and pollution often are denoted by economists as externalities, implying that these are not accounted for in economic models in terms of costs; on the contrary, e.g. in case of an environmental disaster, the additional work returns as a benefit. A correct incorporation of these externalities in economic systems is a key element in bringing sustainability within the economy. [9,51,58].
- An example to move harmful emissions in economic models is the emission trade of CO_2 in Europe [48,44,64]. The government sets a maximum amount of CO_2 emission, which is the basis of a trade mechanism of rights to emit CO_2. A key element in a successful application is that government strictly adheres to a maximum limit, and is not seduced to be generous in dividing CO_2 rights.

---

of people [44,78] p. 369, have a key role in providing products to foresee in needs, and they also are involved in large-scale material handling and energy usage.

At the start of the second decade of the 21st century, the mechanism to work on sustainability mainly is on a voluntary basis, because a coercive mechanism by governments pressing companies is absent [79]. However, other mechanisms appear to be strong drivers in this field: it is known that companies follow each other in main trends, and also norms from NGOs and governments appear to be leading for other branches [79]. Business is seen to have no choice but to embrace sustainability [4], accomplished by leaders who, by vision or from pragmatic reasons, want to keep up [5], or who feel pressed by NGOs reporting on social or environmental burdens of production processes [44].

Many leading experts therefore foresee that sustainability will be key in future and will be part of the business strategy [4,69,80,81]. With first signs

---

BOX 3.5

EXAMPLES OF BUSINESS AND SUSTAINABILITY

A well-known example of a pioneer in creating a sustainable vision is the zero-to-landfill approach of Xerox, manufacturer of copiers. Extensively described elsewhere, shortly, Xerox shifted to a supplier of copier facilities instead of selling equipment only. Realizing that software and some electronic parts usually are limiting the copier's life, rather than the housing, typical contracts of Xerox incorporate regular recycling and upgrading of copiers [2]. Especially industries converting large flows of materials and energy already have many cases available, as reduction of costs is a main driver [129,130]. Examples are DuPont [2,131] and General Electric [2], and also the automotive industry [132].

---

in the end of the 1990s [27], nowadays, many examples of (branches of) industries and companies are known working on sustainability, see Box 3.5. For example, the message of the World Business Council on Sustainable Development (WBCSD) is very clear. WBCSD is a CEO-led, global association of some 200 companies dealing exclusively with business and sustainable development which aims to represent around 1/3 of the total generated GNP [2]. The claim of WBSCD is that business wants and needs to have the lead in the pathway in future, including poverty, climate change, resource depletion, globalization, and demographic shifts [69].

Implementing sustainability in business practice may feel not only for the manager, but also for the engineer, as of improving quality [5]. The total quality management of e.g. Toyota, not only seemed a good internal driver for improving production processes, but also has a strong marketing value in brand and reputation [5]. Starting with incremental changes in production processes and products, the foreseen long-term objective is a radical change [5,34]. Different examples of companies are available, which are on this pathway to sustainability, characterized by stages – 'Do old things in new ways' (e.g. prevention of pollution of 3M), 'Do new things in new ways' (e.g. zero waste commitment of DuPont), 'Transform core business' (e.g. incorporate technology breakthrough by Dow), and 'New business model creation and differentiation' (reposition the company as GE) [5]. Text Box 3.6 shows other representations of pathways. All these business models are reported to provide a useful framework to work on sustainability.

The sustainability value framework provides a clear representation of business strategies [2,82], Fig. 3.11. The lower left quadrant is the most

BOX 3.6

TYPICAL BUSINESS STRATEGY STAGES IN SUSTAINABILITY

More differentiated to sustainability, typical stages are [4]:

'Viewing Compliance as Opportunity',
'Making Value Chains Sustainable',
'Designing Sustainable Products and Services',
'Developing New Business Models', and
'Creating Next-Practice Platforms'.

Stages in the framework of culture change [58]:

awakening: defining the vision;
cocooning: creating the road map;
metamorphosis: aligning the organization;
emergence: ongoing integration; and
engagement: influencing the others.

FIGURE 3.11 Sustainability value framework [2,27]. Matrix inner box: Corporate payoff; matrix outer boxes: Strategy. Left and right arrow type of boxes: drivers.

defensive one, in reducing cost and risk of the internal processes. The relation with the outside in the lower right corner is to establish the license to operate: reputation and legitimacy. More proactive is the upper half of the matrix: developing clean technology (right upper corner) gives the opportunity to repositioning the firm. Introducing a sustainability vision in the left upper quadrant is a recipe of a growth trajectory in which a shared road map with society aims to address general needs, as poverty and equity. Therefore, not only the environment is involved, as inequity [53] is playing a significant role in world business trading.

A large share of executives now see sustainability as a revenue driver [4,5,81]. The green marketing is doing an effective job, with an increase of 500% from 2005 to 2007 of eco-friendly product launches, thanks to an aggressive leadership of the world's biggest companies [81]. Many enterprises are on pathways to a change, as can be easily verified in browsing leading business journals (see Box 3.5).

Different formats are available as concrete instruments to incorporate sustainability. Corporations have their business goal and cannot be expected to fully act as society might wish or demand. Instruments help to line up business strategy. Corporate social responsibility (CSR) [80,83] is an example of an important instrument [84], to handle sense making [84] connected to moral obligation, sustainability, license to operate, or reputation [80]. The UN Global Compact is looked upon as a serious direction to CSR [79] providing guidelines which ask companies to embrace, support, and enact, within their sphere of influence, a set of core values in the areas of human rights, labor standards, the environment, and anti-corruption [85], see Box 3.7.

Corporate sustainability usually is directed to a specific field of attention [86]: on the business case, the natural case or the societal case, and intermediates. Figure 3.12 structures these cases with six criteria [86]. A company is directed to the business case, but may move to a strategy in which the natural (ecology) or the societal case becomes substantial. Popularized by the WBSCD, eco-efficiency is a well-known term to bring sustainability to the business context. It is mainly directed to reduce the environmental impact of goods and services. However, the starting point is to adhere at current products and processes. But, at increasing production, even at a high level of eco-efficiency, the absolute number of pollution or resource depletion may increase as well. In case of, e.g. manufacturing petrol-fueled cars, the natural resources are not capable of delivering of materials and energy when production rates are increasing. Eco-effectiveness is a more fundamental shift to the natural case, e.g. in producing electric vehicles. The same reasoning holds for the societal case: socio-efficiency is mainly directed to the internal processes, while

---

BOX 3.7

UN GLOBAL COMPACT PRINCIPLES

I. Human Rights;
  1. Businesses should support and respect the protection of internationally proclaimed human rights; and
  2. Make sure that they are not complicit in human rights abuses.
II. Labor
  3. Businesses should uphold the freedom of association and the effective recognition of the right to collective bargaining;
  4. The elimination of all forms of forced and compulsory labor;
  5. The effective abolition of child labor; and
  6. The elimination of discrimination in respect of employment and occupation.
III. Environment
  7. Businesses are asked to support a precautionary approach to environmental challenges;
  8. Undertake initiatives to promote greater environmental responsibility; and
  9. Encourage the development and diffusion of environmentally friendly technologies.
IV. Anti-corruption
  10. Businesses should work against corruption in all its forms, including extortion and bribery.

Source: [85]

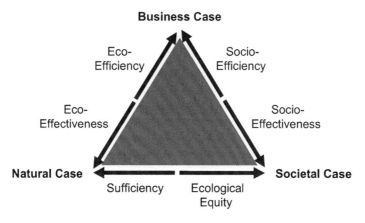

FIGURE 3.12 Structuring corporate sustainability [86].

socio-effectively is involved in societal needs, e.g. a pharmaceutical company providing products for poor countries [86]. Companies already operating in the natural or societal case mode may shift to sufficiency or ecological equity.

A term less directed to one of the corners of Fig. 3.12 is eco-innovation, which contains both environments as (other) stakeholder requirements. Eco-innovation is focused on the business side of designing products and services in which not only the user but also governance is central [6,37]. Business use these terms to embed sustainability in their business strategy, connected to procedures or pathways, of which eco-innovation, eco-efficiency, and sufficiency are most common [21,22,37,54]. Despite the terms, the values of business are connected to its roots in society; business is dependent on a good social structure, as education, political stability, and infrastructure.

Engineers are likely to expect to work in business which is facing the challenges, opportunities, and difficulties of sustainability. As this section shows, business is seriously incorporating sustainability. Large transnational companies may aim a high position in rankings based on sustainability performance indicators, as the Dow Jones Sustainability Index, see Box 3.8. Reporting by GRI (Global Reporting Initiative, of the UN, [83] p. 180) may induce procedures on sustainability throughout the whole company. Marketing departments embrace the foreseen revenues of having an eco-premium [5,87].

Despite all these drivers, from global, society, and business, it is the engineer working alone or in a team who supports or obstructs the working on sustainability, which is the topic of the next section.

BOX 3.8

(DOW JONES) SUSTAINABILITY INDEX

The DJSI family, as an well known example of sustainability indexes, uses a best-in-class approach to selecting sustainability leaders from all industry sectors on the basis of defined sustainability criteria embedded in the SAM Corporate Sustainability Assessment. It comprises an in-depth analysis of the world's largest companies based on economic, environmental, and social criteria, such as corporate conduct, labor practices, and environmental policies, with a special focus on industry-specific risks and opportunities that companies face.

Source: http://www.sustainability-indexes.com/

CONTEXT LEVEL ENGINEERS

Sustainability adds requirements, specification, and challenges to the design assignment of the engineer. The importance of understanding context for designing in general is familiar to the engineer. An artifact (something created by humans) should be useful, safe, and fit for a specific context in which it is used. In that respect, an engineer is used to assess the design on norms related to the local situation. Many of these norms are settled in prescriptions on the design, made by government and work field associations. For example, for industrial equipment, the usage of the BAT (best available techniques) is in most cases required to get a permit related to the environment. Other possibilities are more fundamental approaches such as green chemistry or cradle to cradle, see Box 3.9. During the last few decades, the safety regulations have become very strict, with detailed conditions according to labor force and the surroundings. From the perspective of the engineer, as a designer, sustainability adds a new type of conditions to the design assignment.

BOX 3.9

CLOSING THE CYCLE

On a more fundamental scale, the principle of green chemistry provides a means of a fundamental reduction of waste streams, e.g. by performing addition reaction instead of substitution reaction [133]. The cradle-to-cradle principle advocates to even abandon waste (waste equals food), which is a challenge for chemists to establish a true recycle flow (instead of downcycling) [7].

The largest differences between sustainability and e.g. safety is that the latter one is far better regulated and deals with relative short-term local effects, while sustainability deals with long-term global effects. Regulations prevent realization of a sloppy design, in which safety requirements are not well met. For sustainability, strict regulations are not (yet) available. However, the freedom to choose the extent of a sustainable design is rather restricted, for several reasons: the responsibility of the professional, norms of society, personal and professional ethics, working in a team, and the business strategy and challenge.

Responsibility of the Professional

The increasing knowledge on sustainability should be familiar to the designer. The designer has an obligation to make a determination regarding public welfare, health, safety, and environment [15], p. 118. It is the standard of the designer which determines the quality of the design, although tested and reviewed by users and stakeholders. Similar as to the quality issue, as a professional, the quality of a design should meet standards of care; thus, it should be at least safe [15], p. 103. The same standard of care demands that the designer should be aware of the development of new knowledge to anticipate on developments in a broad sense, also in the field of sustainability.

The standard is that, in general, the design decisions on sustainability should be consistent with what a reasonable person should do in similar circumstances [15], p. 114. In e.g. housing, building prescriptions nowadays usually demand a high insulation value. The designer can be expected to have knowledge on the different types of insulation techniques, and their pros and cons. In case of energy-saving projects in industry, the standard of requirements is directed by money saving. A professional on cooling water systems may be expected to be aware of aspects of material usage, fouling, high- and low-quality heat, and so on. As thinking on sustainability is on a substantial level, and many practical cases are available, for the typical field of interest, the professional should be aware what the latest techniques regarding sustainability are.

Norms from Society

What is reasonable to society is indicated and directed by norms. 'Reasonable' implies that the designer imagines what the implications of his design are for society and that he is obliged to apply society norms to his design, such that the implications will be acceptable. This obligation has of course its limits [15], p. 118. Norms are, similar to morality, not in the form of one set of very specified and strict rules, but may differ with place and in time [15], p. 49. A normative discussion is concerned with rules and principles to govern our thoughts and actions [44], p. 34, to provide basically principles for distinguishing right from wrong [15], p. 47. Therefore, a good assessment of the choices in the design process requires a good feeling with the context in which the design outcome will act. The final choices are usually a compromise of many influences, as the engineer may be the leader of the design assignment; he or she is not necessarily the sole decision maker [15], p. 111.

Drivers from a global context level are codes from engineering associations, and internal directions from strategic management of global companies. These drivers all add to the decisions to be made in design assignments. However, it is the individual engineer, or, in most cases, the design team (of engineers) that takes design decisions [15]. It is the designer or the design team that puts emphasis on certain design aspects, and decides e.g. to try to incorporate renewable materials. What are their obligations? Are they completely free to act, within the boundaries set by the design assignment or not? The field of ethics provides a structure which separates the responsibilities of the professional engineer and the engineer as person. Also, a distinction is made between the work in a team and on an individual basis. Knowledge of the fundamentals of this structuring will be of help in making decisions.

Professional and Personal Ethics

Ethical theories make a distinction between professional ethics and personal values [50,83]. Personal ethics are far more institutionalized and to a certain extent regulated by law and common morality. However, the professional engineer also has obligations to act, because it is expected that after a thorough education, the professional is capable of performing well in the field of the discipline. In general, six qualities are needed [83]:

1. Integrity and openness and honesty, both with yourself and with others
2. Independence, to be free of secondary interests with other parties
3. Impartiality, to be free of bias and unbalanced interests
4. Responsibility, the recognition and acceptance of personal commitment
5. Competence, a thorough knowledge of the work you undertake to do
6. Discretion, care with communications, and trustworthiness.

As these qualities are generally shared by all professional practitioners, see e.g. on safety of Box 3.10, they may give an awareness in daily work life, by which a decision is more easily made and maintained, in combination with personal values [83].

Decisions in sustainable design incorporate values which are an assessment or measure of the worth of something. Ethics make values explicit. Normative ethics are on questions of 'ought', with multiple and conflicting values which are used by stakeholders to make judgments on the realities they experience [50], p. 161. The perspective is of importance, e.g., the anthropocentric or ecocentric perspective in environmental ethics [50], p. 161. There is no such thing as a final solution — values and perspectives vary and develop according to the context and time in which they are applied. Therefore, ethics are like design in that there are multiple good solutions to a design problem, both from a technical and from a moral point of view — this does not mean that all solutions are good ones [88].

BOX 3.10

SAFETY

Professional ethics related to responsibility on safety have a longer history and are evaluated to the following categorization [89]:

- minimalist, engineers are bound to standard operating procedures, to avoid blame and liability — this level is commonly applied for most companies;
- reasonable care: the engineer provides protection to society, based on a risk analysis of the technology applied; and
- good works: potential hazards are examined and additional steps beyond the expected risk analysis are taken to safeguard against the hazards.

Team Role

Working in a team implies that the focus is more on social ethics and less on individual ethics of persons [47]. Most engineers work in teams, which are most of the time project teams on an ad-hoc basis. What is the responsibility of a team, compared to the individual? From an ethical point of view, the responsibility is clear: a team feels responsible as a team and the individual members feel bound to their personal and professional code and ethics. In teamwork, group processes do play an important role, which adds another dimension to the ethical considerations. Individual ethics usually do not properly address this type of social implications, because they mainly focus on the integrity of the individual [47]. A discussion between team members clarifying the values and norms to work by during the design project will help to bring the team ethics closer to the individual ethics. A final answer on regulating group processes in teamwork is not available, which leaves that the engineer should be aware that these group processes require explicit attention in making assessments related to sustainability.

Engineering Codes

Engineering codes for engineers, provided by engineering associations and institutes, give a centralized norm on the moral handling of engineers. The standards of care are leading, but not exactly defined, as they should be sensibly applied to specific local situations [15], p. 67. In different circumstances, in different times and regions, the interpretation of care is different. In specific cases, associations may come up with technical guidelines, such as green chemistry. The guidelines from the engineering

---

BOX 3.11

GENERAL SETUP OF PROFESSIONAL ETHICAL CODES [83]

1. Responsibility to the profession.
2. Responsibility to oneself.
3. Responsibility to the employer, with the member acting as an employee.
4. Responsibility to the client.
5. Responsibility to the other individual members of the group or profession.
6. Responsibility to the community.
7. Responsibility to the environment.
8. Responsibility to other groups of professions.
9. Issues concerning responsibilities of confidentially and probably cover whistleblowing.
10. Statements on if members have broken the institution's ethical guidelines.

---

associations are derived from general human values and ethical judgments. Box 3.11 shows the general setup of professional ethical codes.

The risk assessment and possibilities to mitigate those risks by engineers may be extended into norms and are rubricated in engineering codes. However, the engineering professions are not yet bound by them, in contrast to licensed practitioners such as solicitors and medical doctors who are bound to uphold their profession's formal ethical code [88]. As registered engineers are coming up, the function of these codes may change from voluntary to mandatory guidelines. Various branches of engineering societies have developed their own set of codes. Comparing the codes, they all share a relatively small set of central values, among which the most important are [83,88]:

- overall mission of the profession, as contributing to human welfare;
- in line with the previous one, the large importance of public safety, health and welfare, and protection of the environment;
- responsibility to be competent in one's work, and the need for ongoing professional development;
- loyalty to both employer and clients or customers; and
- related to fairness, respect intellectual property of others and avoid conflicts of interest, discrimination, and unfair competition (e.g. hire employers of competitive companies).

These values are increasingly related to globalization, which e.g. results in a more international oriented work field for engineers. It should be noted that the codes are usually focused on humanity only, see Box 3.11. High-quality work in engineering includes technical, ethical, and competences as communications and teamwork skills [88]. One of the exceptions is the World Federation of Engineering Organizations (WFEO) which has included requirements ("shall") on avoidance of resource depletion, and creating healthy surroundings [89].

Challenges for the engineer

Apart from what is the responsibility, what are norms in society, and ethical considerations, the challenge to apply sustainable business challenges to the design can be very stimulating. Inspiring leaders and movements show vistas to new processes and products. Examples ranging from large, to medium and small enterprises, and from government and NGOs, show that working on sustainability is moving many people to rethink processes and products. Striving on a high ranking on sustainability lists, such as the Dow Jones Sustainability Index (DJSI), requires a thorough review of existing processes and products. Incremental changes usually are not enough, but statements are needed as carbon neutral, or zero-to-landfill, in case of Xerox already many years ago. A glance at the various lists of the DJSI reveals that many of the companies mentioned are popular employers; in general, employees highly value substantial efforts on sustainability of their employer. Working as an engineer within this atmosphere may be expected to be inspiring and motivating.

Now that the sustainable context in which the sustainable design has to be embedded has been identified, the generation of design solutions can be started as described in chapter 4.

4

Creating Design Solutions

This chapter describes the procedure and specific tools to facilitate the creation of (sustainable) design solutions. Tools are specifically selected and developed for the ideation stage of the innovation funnel and for concept design.

Engineering for Sustainability,
DOI: 10.1016/B978-0-444-53846-8.00004-4

61

The section 'Design Synthesis' provides generic applicable tools concerning the synthesis of the design. Elements are the concept of process flow sheets, how to generate ideas by brainstorm sessions, and how to establish effective teamwork. Risk assessment and mitigation measures are here also placed, as they can be used in the problem definition stage to decide on the degree of novelty and the related research and development effort allowed by the company, as described in chapter 2.

The next section ('Preliminary Solutions Assessment') is an introduction on the analysis part of the design cycle, followed by a section on a Quick Scan Life Cycle Analysis, which can be used in problem definition phase and in the analysis step of the concept design process. Finally, section 'Evaluation of the Design' provides rapid analyses for the concept developed in the idea stage and tools for the evaluation of the concept design results, namely building a set of scenarios and red flag method.

Figure 4.1 gives an overview of the contents of chapter 4. The three steps after the problem statement (refer to Chapter 1) are synthesis, analysis, and evaluation, which all are applied in a recycle loop to optimize the design solution. General applicable steps, that are steps which are not specifically for a sustainable design, are on the left side, such as function design, representation, etc. Typical tools for facilitating a sustainable design are on the right side, such as life cycle assessment, quick evaluation tools for economic and social acceptance, and scenario set building.

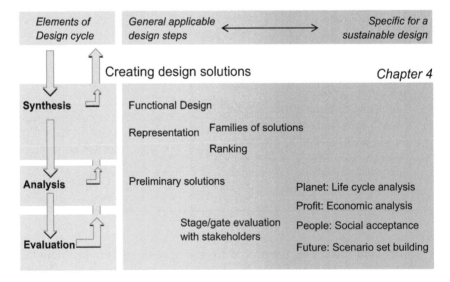

FIGURE 4.1 Subpart of Fig. 1.2, denoting the contents of chapter 4: Creating design solutions by systematic application of the synthesis–analysis–evaluation steps.

DESIGN SYNTHESIS

In the design synthesis step, solutions are generated. Creativity plays an important role here. This means that a simple step-by-step method to generate solutions is impossible to define [90]. For synthesizing solutions, therefore, only hints and refer-to methods and tools are described. The designer should feel free to generate first preliminary and incomplete solutions. By analyzing these, ideas for improvement will pop up, or additional information will be gathered, or the problem will be redefined. In this trial-and-error way, the final design result will be obtained.

This paragraph provides some practical methods in the design synthesis steps: representation of the design, brainstorm sessions, and grouping of functions.

Integral Synthesis

Very good designs solve several problems in one integral way. This integral solution starts with the desire to have a far better sustainable solution than present conventional solutions.

Here is an example of a sustainable building. It is an office and laboratory building for the Wuhan New Energy Institute, Shanghai, China. It is presently under construction and will be delivered in 2012. The building has the shape of a mushroom and is 135 meters high. The top part is much wider than the stem-shaped bottom part of the building. The wide top provides shade for the bottom part to keep it cool. The slightly tilted top part contains solar cells for electricity production. The stem part creates natural convection so that no vans are needed for air conditioning. The stem is also used for wind electricity production by having windmills with vertical axes. In the rim of the top part, tap water is heated. Furthermore, rainwater from the top part is collected and used for toilet flushing, for feeding cooling towers, and for irrigation of the gardens surrounding the building. The building will be evaluated by the British BRE and probably will get the highest BREEAM certificate and becomes the most sustainable building in the world. The building is also pleasing to the eye.

Victor Pastoor, manager of technical specialists at Grondmij, engineering contractor firm for the building, states that normally a sustainable building is obtained by a conventional design and then additional technologies for energy savings are added at the end. Here, the design started with a sustainability vision and that created the integral solution [91].

In many engineering disciplines, it is a design practice to design first using functions only and later designing by defining the form and the

---

BOX 4.1

AN EXAMPLE OF FUNCTION ANALYSIS IN THE PROCESS SYNTHESIS

A function defines what needs to be done, and not how it should be done. An example for process synthesis illustrates its application. Functions in the process industry are reaction, concentration change, separation, and change of phase, temperature, pressure, and form. Such a functional design can be quickly made starting from the input streams and seeing what is needed to come to the output streams.

After the functions have been defined, an attempt should be made to combine functions into a feasible process unit operation. Examples are reactive distillation, reactive extraction, and reactive extrusion. Methods for function identification and integration are described in detail by Siirola [134]. Function integration is part of a novel way of designing processes called process intensification (PI). A good book introducing PI and also giving industrial cases is provided by [135].

---

materials. This is in particular useful for radically novel designs as in general are needed for a sustainable design. Box 4.1 shows how this is done for process synthesis. For product design and service design, however, similar steps are to be taken. First, the concept designer focuses on what is needed (defining the functions) and later how these necessary functions can be integrated into a real design with a shape and specific materials.

Generation of Solutions

In a sustainable conceptual innovation, the design has to be communicated to various stakeholders. This means that the design should be represented in a clear and understandable way. For process concept designs these are simple block flow schemes showing the process steps as blocks and streams as arrows. The following rules help to make the flow sheet very easy to read:

- Place most blocks in a sequence connected by stream arrows pointing from left to right. Some blocks can be placed below other blocks connected by vertical stream arrows.
- Give each block meaningful names.

- Show recycle streams by pointing them all from right to left.
- Keep recycle streams empty; hence, do not place blocks in recycle streams.
- Give each stream a number S1, S2, and provide a stream table showing the contents of each stream.

For products, the representation should be pictures. The pictures should show the product functions and the product components. As in product design, the composition of the product should also be shown in composition table in terms of materials and the amount involved.

Animation films are also helpful for complex products and for products such as buildings, ships, and airplanes that are so large that they affect the horizon view.

Brainstorm sessions in groups can be helpful in generating a large amount of ideas. It can be used at the start of an innovation project to obtain different potential concepts. It can also be used to solve a particular problem later in the innovation stages. A method for organizing a brainstorm session and how to handle the brainstorm result afterward is found in chapter 2. Many ideas can be obtained by using the method of "crowd surfing" on the Internet. Via specific Internet sites the design problem can be stated and people from all over the world provide solutions for free. A book on details of the method is by [92]

Generating **different families of solutions** also helps to come to good designs. Different families can be generated by applying different technical principles for solving the problem. In a process design, one family may be based on homogeneous catalysis and another on heterogeneous catalysis. In a product design of, for instance, a chewing gum that does not foul the streets, it can be that the gum is degraded in the mouth and is swallowed or on the principle that when landing on the street it does not stick and degrades quickly when washed with water. The different solutions can then each be strengthened by using part solutions of one design and applying it to the other design. Working with families of solutions will also reduce the chance of having a designer's block similar to a writer's block.

Ranking the different intermediate solutions and selecting the top solution for further work out can help. A structural procedure for industry is given in Box 4.1, applying process synthesis using functions. If in a later stage that solution appears not fulfilling all mandatory criteria, then the next best solution can be taken for further work out.

Risk Assessment and Mitigation by Research and Development

Risk assessment of a new process or product design is not easy. First of all the term 'new' has to be defined. "New" means not applied at a commercial scale for that particular application. So, for instance, a filter

TABLE 4.1   Risk Assessment Table [143].

	New equipment	New chemistry	New product	New market
New equipment	Low–Medium	Medium	High	High
New chemistry	Medium	Low	High	High
New product	High	High	Low–Medium	High
New market	High	High	High	Low

already used for a particular application A and now for the first time used for application B has to be classified as new and an appropriate development effort has to be followed to mitigate the risk of failure.

Classification of risk categories will help to assess the risk. The categories are new equipment, new chemistry, new products, and new market. If the new concept falls in only one category, the risk can in general be assessed and mitigated by research and development. The risk assessment Table 4.1 can be used to assess the risk level. Box 4.6 gives an idea of the red flag method, used in practice, for a quick evaluation by applying stakeholders.

For a new process for an existing bulk product, in general, a pilot plant development effort will be needed to reduce the risk to an acceptable level. The pilot plant should be a downscaled version of the commercial-scale plant, having all critical features, such as all recycle flows.

If the new process involves novel equipment, then scale-up knowledge has to be generated by research and development; otherwise, the risk of failure for the commercial-scale process will be too large. This scale-up is an important and large subject, which cannot be treated in this book. Various scale-up methods for various types of process equipment are however provided by [93,94,95].

If the product is a performance product, then also the product produced by the new process will in general require pilot testing because trace amounts in the product from the new process can be different from the product made in the old process and changes in the trace composition can change the product performance. For new products (using proven processes and existing markets), market surveys and market piloting will be needed to mitigate the risk.

Case: Design and Development of Industrial Ecology System

Industrial ecology systems, also called industrial symbiosis systems and eco-parks, are important ways of reducing the impact of processes on the environment and on reducing the input of resources. In general, these systems are not designed from scratch and built in one go, but they are slowly growing by the addition of processes, who can use a waste outlet of a certain process and/or produce an intermediate product or utility (steam, electricity, and clean water) that can be used as an input to another process. This gradual improvement of an industrial park is the same method as is used for improving an individual process. It is called revamping and is often combined with increasing the production capacity called de-bottlenecking.

This revamping and de-bottlenecking is, in view of sustainable development, a very important one, because it increases the utilization of existing capital, reduces energy requirements per ton, and, even more importantly, it is an important way of changing a waste stream into a product stream. Also for new products, a revamp of an existing process to produce this new product is to be considered.

Revamps are modifications of existing processes. Revamps are applicable to operations, equipment operation conditions, and also to input and output changes – applicable in all process industries. For novel products, often existing processes can be applied by small modifications. It is often economically very attractive. It also improves the energy efficiency, because the existing utilities are more utilized and becomes a smaller fraction of the total energy required; hence, the energy per ton is reduced.

Here, the term maintenance energy is very useful. It is defined as the energy required for the process to remain at operational conditions even if no product is produced, hence for keeping the process equipment warm and utilities available. The amount of maintenance energy can easily be 20% of the total energy. If now by a revamp the production capacity of the process is increased, this maintenance energy becomes a smaller fraction of the total energy required and thereby the energy per ton is reduced. Sometimes the de-bottlenecking revamp is combined with increased energy integration, by which the primary energy requirement is further reduced.

The author has worked on a process where the initial capacity of the process in 1974 has increased in 30 years by several revamps to 300% of the initial capacity. Hence, revamps can save energy in significant amounts.

However, revamp requires an extensive problem definition, revealing all constraints of the existing plants and therefore an intimate knowledge of the existing plant is needed. Changes in industrial parks should also be treated as revamps. The Kalundburg industrial park of Fig. 4.2 has developed over decades into the sustainable complex it now is.

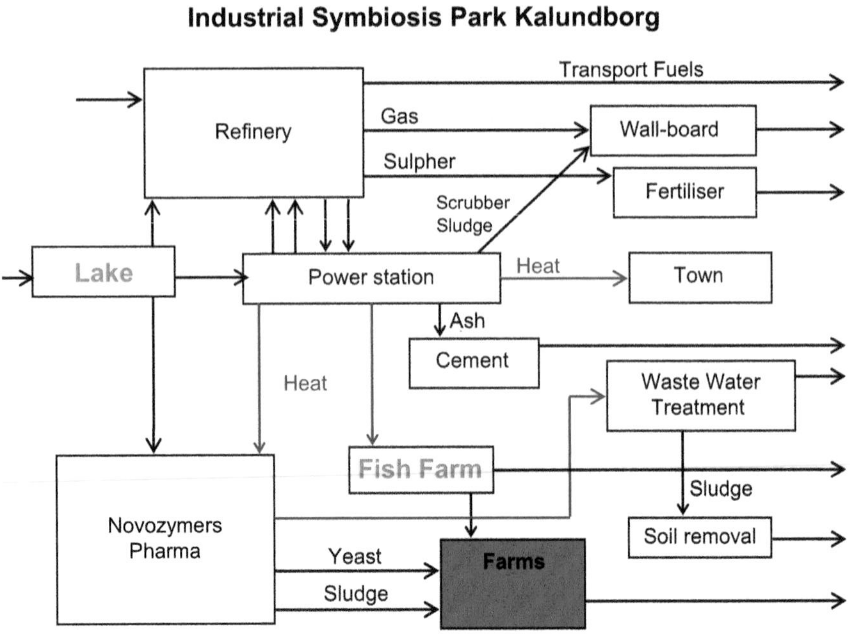

Industrial Symbiosis Park Kalundborg

FIGURE 4.2   Industrial symbiosis park Kalundburg [141].

PRELIMINARY SOLUTIONS ASSESSMENT

Preliminary solutions generated in the synthesis step need to be analyzed on meeting the goals and constrains. For process designs, this analysis is nowadays mostly carried out by computer simulations, called flow sheet programs. For industrial products, a 3-D simulation can be used to find out whether the design meets the objectives and constraints. For complex products, a rapidly built prototype is also a way of finding out whether the design meets the goals and constraints.

The preliminary design should be quickly tested on meeting the sustainable goals and constraints. Quick scan sustainable development tools and methods specifically developed by the authors are described in this chapter 4 and can be used for this rapid testing.

The result in the analysis stage can be even more simply checked by using the simple checklist of Box 4.2. It is an example of a quick and simple checklist for doing a first sustainability analysis for any industry branch.

If the solution fails then the design should be improved. If it is unclear in what direction the design parameters should be changed to obtain a good result, then an experimental design scheme, like a full factorial

BOX 4.2

SUSTAINABILITY ANALYSIS LIST FOR INDUSTRY

Goal

Does it provide in a need (enhancing human welfare)?

Economic

Is it profitable (short term, long term)?

Does it not deplete a scarce resource (long term, worldwide)?

Does it not cause external cost?

Social

Is this technology in compliance with government regulation?

Does it fit with local society (safety, health, and culture)?

No suspected toxic components in the product?

Environmental

No suspected harmful emissions to atmosphere, water or soil process, and life cycle?

design for the parameters, should be drawn up and the simulation program predicting the performance of the process or the product should be run for all defined parameter values, so that the desired solution is found. The rapid sustainable product assessment method of Section 4.4 is a useful tool to analyze the product design on the goals and constraints and how it compares to an existing reference product. If various design solutions have been generated, then they should be ranked and the best chosen.

QUICK SCAN LIFE CYCLE ASSESSMENT

Set-up of LCA

Life cycle assessment (LCA) is a method for calculating the environmental impact of a product or service. The basis for calculation is the so-called 'functional unit'. This may be a unit of material (e.g. a kg of steel of given composition and quality), a unit of energy (e.g. a kW hour of electricity), or a unit of service (e.g. packaging one liter of milk). The basic idea of LCA is that the analysis is done over the entire 'life cycle' of the product or service (see Fig. 4.3). Thus, not only the production phase (typically the domain of the engineer), but also all phases: pre-manufacture, manufacture, use, and disposal of the product,

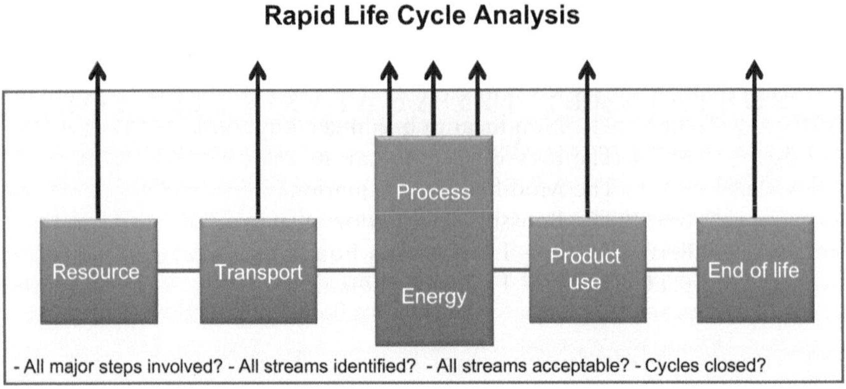

Rapid Life Cycle Analysis

FIGURE 4.3   Example of steps in life cycle analysis.

including all relevant infrastructure (e.g. the plant to make the product and its decommissioning).

A full life cycle assessment takes, in general, months to complete even for an experienced professional. The reason for this is that a lot of information has to be gathered and the quality has to be assessed. To ensure the quality of a formal life cycle analysis, it is very important that the ISO 14040 method is applied; otherwise, the results can be meaningless and the conclusions drawn can be wrong [96].

In a conceptual design stage, this is far too long. The designer wants a first indication in a few hours, for which a rapid LCA method as described here works fairly well. Life cycle assessment (LCA) consists of the following stages, which are briefly discussed and remarks on the quick scan method are inserted. The quick scan LCA steps are summarized in Box 4.3

Goal Definition and Scoping

It is very important to first set the **goal** of the life cycle analysis or assessment. In the conceptual design stage, the goal in general will be identifying the major environmental impacts of the reference process and showing how the new design reduces these impacts (while keeping the other impacts low).

The scope of the design should be clearly distinguished from the scope of the LCA. Within the design object scope, the designer has freedom of designing. The LCA scope however is much wider than the design object scope and contains steps upfront of the design and downstream of the design, which are not changed in technology by the designer. The designer however can affect the emissions of the life cycle anywhere by his design choices, by which changes in inputs and outputs

---

BOX 4.3

QUICK SCAN LCA

1. Goal definition and scoping
>   Define functional unit.
>   Define system boundaries.

2. Inventory
>   Define all life cycle steps.
>   Draw all input and output streams (air, water, and soil).
>   Determine key components in each stream.
>   Quantify key components of streams.

3. Impact assessment
>   Determine types of pollution.
>   Determine sizes of pollution.

4. Valuation
>   Normative criterion for valuation: The new design should be better
>   in some emission and input types and not worse on any of the
>   other emission and input types.

5. Improvement
>   Identify major contributions to pollutions.
>   Reduce pollution by re-design relevant step.

---

of the design are changed, and thereby the life cycle emissions are changed.

For concept design, the LCA scope can be limited to only the major emission stream and also the accuracy is allowed to be low. For factor 4 improvements in environmental impacts 50% inaccuracy is still acceptable.

Part of the goal definition is settling the **functional unit**, which for product design and for large systems design, like a city transportation system, is very important. The way the functional unit will be defined determines what alternatives will be taken into account and will decide what will be designed. For process designs, the functional unit will in general be the design product capacity.

The **system boundaries** of the LCA should be from primary resources to end of cycle (cradle to grave) and preferably include upgrading of the end of cycle output stream (cradle to cradle). For processes, both the feedstock and product stream connections as the main process construction materials should be taken into account and also from primary resources to end of cycle. Classic examples showing the importance of this are nuclear power plants, where the end of cycle of the waste fuel and the power plant itself are causing the major impact on the environment (and on the cost).

For making a political (government) policy, at least cradle-to-grave system boundaries should be defined, but even better are cradle-to-cradle boundaries.

An example of the strength of such a policy system design is for dioxins in the Netherlands. The LCA study on PVC and dioxins (which have extremely high human and eco-toxicities) revealed that neither the PVC production nor use was the problem, but it was their eventual inefficient and 'dirty' incineration in municipal waste-treating plants, where dioxins where formed and emitted via the gas effluent to the atmosphere. Therefore, the (then) most efficient method of reducing these impacts was to improve the performance of such waste-treating plants. This was done in the Netherlands (over the period 1989–2000, by means of more efficient combustion and use of activated carbon filters), resulting in a hundred-fold reduction in dioxin emissions in the country.

Inventory Analysis

The inventory stage determines the **material inputs and outputs** relating to one functional unit of product or service and links these to environmental impacts. This stage is simply a material balance, albeit a complete one. It covers all life steps of the product or service being considered (Fig. 4.3) and defines all inputs and outputs of all of these life cycles.

For the quick scan life cycle analysis to be used in the concept design stage this is too time consuming. Here are some simple steps to obtain the inventory for concept design.

1. Define only all **major steps** of the life cycle for the reference case and the new design. A reference case is essential for the rapid LCA.
   Figure 4.3 may help in identifying the steps.
2. Draw all **major input and output** streams (to air, water, and soil).

Even if the precise composition or size of certain streams is not known, still the streams should be shown. In this way, these streams can be discussed and crude estimates of their composition and size can be made in combination with a sensitivity analysis to determine whether the estimate is good enough for making certain design decisions and for comparing the design with the reference case.

3. Determine **key components** in each stream.

For streams to the atmosphere, this will often be carbon dioxide. For streams to surface water, these will often be inorganics, as these will remain in the water and can affect the water ecology system.

4. Quantify key components of streams.

For a quick scan LCA for concept designs, often a crudely estimated quantification of one key component is sufficient to show the major impacts and make a meaningful comparison between the concept design and the reference case possible.

Impact Assessment

The material and energy flows of the LCA are in the stage of impact assessment both qualitatively and quantitatively evaluated on their contribution to pollution and the size.

Generally recognized **environmental impacts** are given in the Table 4.2. Assigning an emission of a given substance into a given environmental impact depends on the properties of a given substance. Emitted sulfur dioxide, for example, is converted into sulfuric acid and is therefore classified as an 'acidification' impact. Emitted carbon dioxide and nitrous oxide are greenhouse gases and are therefore classified as 'global warming' substances.

Once emissions have been classified into a given environmental impact, they must be assessed. This means calculating the quantitative value of this impact. Environmental impacts are calculated by means of

TABLE 4.2 Generally Recognized Environmental Impacts [94]

Depletion	Pollution	Disturbances
Abiotic resources	Depletion stratospheric ozone layer	Desiccation (dehydration)
Biotic resources	Global warming	Physical ecosystem degradation and landscape degradation
	Formation photochemical oxidants and pollutants	Human victims
	Acidification	
	Human toxicity	
	Eco-toxicity (terrestrial and aquatic)	
	Nutrification (eutrophication)	
	Radiation	
	Thermal pollution (dispersion of heat)	
	Noise	
	Smell	
	Occupational health	

BOX 4.4

GLOBAL WARMING POTENTIAL

Conversion factors for global warming potential (GWP) are expressed in terms of CO_2-equivalency (kg CO_2-equivalent):

Substance	GWP
Carbon dioxide	1
Methane	62
Nitrous oxide	290
Chlorofluorocarbon (CFC)	5000

(GWP = global warming potential = kg CO_2/kg substance)

Using the GWP values, emissions to air of various substances can be converted into an equivalent global warming effect by means of the formula:

$$\text{Global warming effect}(kg) = \sum_{i=1}^{n} (\text{Emissions to air})_i \cdot (GWP)_i$$

Thus, the GWP of a factory that per year emits 1 megaton of CO_2, 10 kilotons of CH_4, and 1 kiloton of N_2O is: $3(1 \cdot 1 \text{ megaton} + 0.01 \cdot 62 + 0.001 \cdot 290) = 1.352$ megaton CO_2-equivalent per year.

conversion factors or so-called equivalency factors. Box 4.4 gives an example for the global warming potential of various gases.

Other environmental impacts can be calculated similarly by using the appropriate equivalency factor. Impacts of human toxicity and eco-toxicity are, for example, determined by factors such as toxicity data (for human, resp. various nonhuman life forms), persistence, bio-accumulation, etc. LCA data and impact factors can be found for specific products from the literature (for chemicals it can be obtained from http://lca.jrc.ec.europa.eu/lcainfohub/datasetArea.vm). Also, a number of software packages and databases are available. A useful package available for free from the Internet is SPIonExcel. It is based on LCA but converts all impacts to required area so that a single output result is obtained. It has specific downloadable packages for fuels, base chemicals, metals, and food and others.

In the quick scan LCA only one or two types of pollution should be considered. The choice should be made by a judgment which types are most significant for the reference case. If, for instance, the design concerns a solution with the goal to reduce global warming gas emissions and carbon dioxide is the major global warming gas of concern then the LCA is focused on the carbon cycle from cradle to cradle. If the design concerns reducing emissions to surface water, then the largest component in the water emission streams should be taken into consideration.

Valuation

The valuations stage is by its very nature normative. When comparing two alternative products, for example, two different spectra of environmental impacts with different values will appear. Is a given value of global warming worse or less bad than a given value of human toxicity? Many attempts are being made and have been made to 'value' results of LCA studies. All of these have some normative basis. Often a panel is used to come to weight factors for the different impacts to obtain a more general value distribution over the various impacts.

In the quick scan LCA method, however, only one pollution type will be considered and by taking the view that the new design should be far better on this type of emission (and assumed to be not worse for the not considered emissions) no real valuation is needed at this concept design stage.

In later stages of the innovation, a full LCA will be carried out to check that indeed the new design has a far lower environmental impact than the reference case and is not worse in other impact types.

Improvement

One of the great uses of LCA is that it can pinpoint places in the life cycle which cause major environmental impacts and thus lead to very efficient improvement of the impact spectrum of a given product. Once the life cycle assessment has been made, it will be clear whether the design meets the environmental goals and hard constraints. The assessment also can be used to further improve the design on the major impacts. The rapid LCA method facilitates several improvements and assessments during the conceptual design.

LCA Epilogue

Life cycle assessment has become a recognized and increasingly widely used method of estimating environmental impacts of products and services. Because the environmental impacts it studies also concern impact types directly related to sustainability (e.g. use of resources, such as fossil fuels and water; climate change; long-term eco-toxicity; etc.), LCA is an extremely important tool in studying the sustainability of various technological developments. A general method and cases are provided by [97,98].

EVALUATION OF THE DESIGN

When the design process has passed the analysis stage, because the design fulfilled, according to the rapid scans, all goals, and constraints, it enters the evaluation stage. Here the design is evaluated by stakeholders outside the design team, such as the internal company clients. For radically novel sustainable designs, it is very useful to evaluate it for robustness to future external influences and show the robustness test to the clients.

In this paragraph, a couple of rapid analysis methods are provided which enables a first check in the analysis stage of the concept design. The three methods presented here, are on economics, social impact, and an integral sustainable development assessment. The presented method has been proven to work in academics courses and in industrial practice.

Stage/Gate Evaluation with Stakeholders

At the end of each stage, the result will be evaluated by stakeholders. The designers will present the design in a meeting in sufficient detail showing both the problem definition and the design results to stakeholder representatives. For a sustainable design, the stakeholders are company business, society, customers, suppliers, and environmental NGOs. For each stakeholder group representatives need to be present in the evaluation meeting. The red flag method (see below) can be used to highlight issues.

An important question is when to invite the stakeholders, external to the company, such as NGOs. In general, this is done at the gate at the end of the development stage, when also the commercial detailed design has been made, because then an enormous amount of information is available and most stakeholder questions can then be answered. However, the disadvantage is that if major objections are raised which make sense, then a re-design and an additional development effort will often be needed. By using the leading and guiding sustainable design principles of this book, the chance of this happening will be smaller; hence, this chance may be taken.

At the meeting or after the meeting a decision will be taken by the company business department whether the stage/gate will be passed into the next gate.

Rapid Economic Analysis Method

Economic assessments of designs in the early stages of innovation require very specific experienced craftspeople, which companies with a large R&D section always have in house. They can make judgments on the likelihood of cost to be added to the design later in the development stage to make the design work. For small-scale companies and for academic people this knowledge is often not available, for whom a simple rapid economic analysis method is provided below.

A rapid economic analysis in the concept stage can be made using the economic potential (EP) calculation method. It is derived from Douglas [99] and modified to suit radically new designs with renewable feedstock. It starts with the simplest calculation and ends with complex calculation. The calculation of EP0 gives a first indication whether the new design is economically viable:

$$EP0 = \text{Sales} - \text{Feedstock cost}$$

The most simple dimensions to use are sales and feedstock cost in \$/year. Sales are obtained by multiplying the process capacity (ton/year) by the sales price (\$/ton). For the sales price the present price can be used, although in future this price will be different, it will be the same for the conventional process and for the new sustainable process. For a comparison, the present price is good enough. The price for the feedstock cost is for conventional fossil based that is simple to obtain from literature and the price is the same globally. For renewable feedstock such as biomass the price is strongly dependent on the transport cost involved.

Therefore, EP1 is introduced, which includes transport cost and EP2 which includes capital cost:

$$EP1 = EP0 - \text{Transport cost}$$
$$EP2 = EP1 - \text{Capital cost charge}$$

Box 4.5 provides an example for the economic evaluation of biomass production.

The capital cost charge can be obtained from capital cost charge=capital investment/payback time.

The payback time depends on the company's policy for economically assessing concepts. Often a payback time of 3 years is set.

More elaborate methods such as net present values will give more accurate results but in general do not change the design decisions to be made in the concept design stage.

BOX 4.5

ECONOMIC EVALUATION OF BIOMASS PRODUCTION

The transport cost to the factory is an important element for renewable resources as feedstock for energy carriers like petrol and diesel. In fact, the transportation costs are linked with the plant capacity of biomass conversion to bio-crude oil: a larger capacity factory means longer transportation distances and vice versa. For fossil crude oil the transportation cost per ton km are far lower because sea transport and pipe land transportation cost are far lower than truck transport over land. Figure 4.4 shows qualitatively the difference between costs of transport and refining for fossil crude oil with costs for biomass. For these processes, the following economic potential calculation may be useful:

EP1 = EP0−transport costs and EP2 = EP1−capital cost.

Rapid Social Acceptance Guideline

Social acceptance of a novel design can only be truly measured by involving people from society and asking their opinion. Companies of consumer goods and services have developed large amounts of

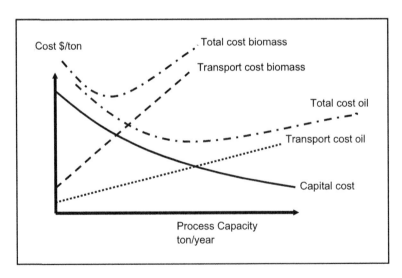

FIGURE 4.4.   Process Capital and Transport Cost Minimalisation

knowledge how to do these surveys. Here, a generic guideline for social acceptance is given.

Social acceptance is strongly linked to human safety and health performance of a process or a product, both perceived and real. So, if the new concept design scores better on safety and health then social acceptance is more likely to occur than a new design that does not have this. This guideline can be applied in the synthesis and analysis concept design cycle.

Rapid Integral Sustainable Development Assessment

Korevaar describes a sustainable development assessment method, which is suited for new products [32](Korevaar, 2004; p. 211–217), and can also be used for new processes. The new design is scored qualitatively by a few expert people from different departments or background (safety, health environmental (SHE), economics, and marketing) on LCA impacts, on social criteria: meeting government regulations, consumer need, social acceptance, human toxicity and human accidents, and on economic criteria. The score is qualitatively from 1 to 5 for both the new design and the reference case. The score of the new design and the reference case is placed in a so-called polar graph as shown in Figure 4.5.

This shows directly where the new design scores less favorably than the existing case. In general, this is unacceptable. The new design should at least be equal in all aspects and score significantly better on several aspects.

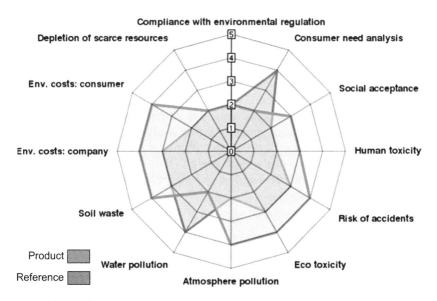

FIGURE 4.5   Polar graph sustainable development assessment [32].

The strength of this method is that it is fast and that various performance aspects of the design can be directly analyzed and compared with the reference case. Only the relative difference to the reference case is important. The weakness is that it is subjective. However, by having a complementary panel this subjectivity is limited. The standard deviation of the panel score was in general sufficiently small to obtain meaningful assessment and outlier results can be discussed to obtain clarification.

This rapid method can also be used for rapid preliminary process design assessment.

Scenario Set Building for Robustness Test to Future Uncertainties

A powerful tool for testing a new design on robustness to future uncertainties is making a set of scenarios. The scenarios are conceivable pictures of the future. Each scenario in the set should be based on different assumptions about the future resulting in that the set of scenarios encompasses all future uncertainties relevant to the design.

If the new design is successful under all scenarios in the set, then the design is considered to be robust to future developments. If the design is not successful under all scenarios, then it is not robust to the future. Often the design can be modified so that it becomes future robust; if not, then the design and development project should stop.

Here are simple steps to generate the set of scenarios and test the design [32] (Korevaar, 2004, p. 191). The steps are written in very short statements, with little explanation.

Step 1: Describe the potential future threats to the design when commercially implemented. Take for the future time span the lifetime span of the designed artifact or service. With life span is meant the foreseen time that the process is in operation or the product or service will be in the market. For processes, this is often 30/50 years. For products, it is typically 1 to 10 years. Threats are uncontrollable events that affect the success of the new design. Threats can be, for instance, feedstock related, product market related, or a breakthrough in a competing technology.
Step 2: Identify all factors for success and failure of the design and determine key trends for each factor. Include also discontinuities – so-called high impact-low probability events.
Step 3: Classify the factors into constant in time, predictable in time, uncertain high impact/low probability.
Step 4: Build the scenarios using the classified factors as elements. Make at least two different scenarios.

BOX 4.6

RED FLAGS METHOD CASE EXAMPLE

An example of stakeholder involvement in plant design is Shell Chemicals Canada, which has a Sustainability Advisory Panel to examine the impacts of its products. The panel developed a scoring system to judge sustainability of products. "Red Flags" has helped Shell Canada to make a major business decision in 1998 to build a new ethylene glycol plant in Scotford.

One of the most interesting results of the panel evaluation was that the carbon dioxide output stream of the plant, originally sent to the atmosphere, was sent to local green houses [136].

Step 5: Make the scenarios visual and give them meaningful names.
Step 6: Apply the scenarios to the design and draw conclusions on the success for each scenario.

N.B. Suppress the temptation to add a probability to each scenario. The strength of the scenario set is that it does not rely on predicting the future. Predicting the future is impossible.

Step 7: Improve the design using the scenario test results of the design.
Step 8 Present the scenario set and the outcome of the robustness test to the stakeholders.

Red Flags Method for Evaluation with Outside Stakeholders

The red flags method is very useful if a large variety of external stakeholders is invited to evaluate the new process or product or both. It means that representatives of all relevant external stakeholders on the environment, society, and business are invited to evaluate the new design on the contribution to sustainable development. Often these stakeholders have little knowledge of the technology and here the red flags come into play. Any time any item of the new design is unclear to the representative, or when he or she disagrees with an attribute of the design, or both he/she raises a red flag and states why he/she raises the red flag. The company then will either try to clarify the item or state what they will do with the criticism. Box 4.6 shows how Shell used this method.

Acquiring Sustainable Design Competences

TEACHING SUSTAINABLE DEVELOPMENT

Acquiring competences to make sustainable design requires knowledge and skills to embed the design into society, economy, and environment. An important aspect of teaching sustainable development is therefore to establish that all three aspects, ecological, economic, and social, of sustainable development are fully incorporated, especially for engineers, whose education is mainly based on reductionist approach [100]. Concentrated on the design result, engineers tend to weigh the outcome of design (consequentialist) [15], p. 48, rather than the underlying rules or motivations of the design. Including social trends and trying to incorporate future views on their design basically are not within their usual design approach. The social part is the most difficult part in framing the design assignments [101], because it is less well defined compared to the other two parts, economic and ecological [102]. A complicating factor in teaching the social aspect of sustainability to engineers is that most (classical) engineering curricula do not have the tradition to incorporate social aspects in their courses [101]. Because engineers are mainly educated to rely on science fields of mathematics, physics, and chemistry [103,104], structural knowledge from the humanities and in particular social aspects usually are not within the framework of engineering programs. This adds to an unfamiliarity of most engineers with the social dimension of their design assignment [103].

Implementing the social part of sustainability is a challenge not only for academic education, but also in a professional work setting. This challenge will however be relevant for more and more engineers, as the social part of sustainability is touching the recent trend that engineers increasingly are becoming decision makers [105,106]. Engineers have to concern matters which are not well defined and more oriented on moral sensitivities, rather than on scientific understanding [13,106]. Sustainability is an illustration of the observation that their work increasingly incorporates the social impact of their design [24].

SUSTAINABILITY COMPLEXITY TRANSFERRED INTO DESIGN COMPETENCES

Education on sustainability is to teach the additional complexity incorporated by making a sustainable design. Box 5.1 shows an example of how the transformation to complex sustainable design has occurred in the field of chemical engineering. This complexity is structured by defining key competences which are essential in making a sustainable design. In this field 'key competences' are commonly denoted as competences which are not connected to a specific subject [107], but holds for general applicable skills. A definite final list on how to (or not) or which competences to apply in is not yet available, but recent reviews [23,107–110] provide a framework in which the course, module, or form of education can be situated and evaluated.

From this, four key competences are defined, derived from the four 'pillars' of the line of Unesco [111] proposed to base education in the 21st century. These 'pillars' shortly are: learning to know, to do, to be, and to live together, or also stated as knowledge, methodological, personal, and social, respectively [111]. The derived key competences are:

1. *The competence to internalize sustainability* ('learning to know', or 'knowledge') in jobs and future career, reflected in the development of a long-term view on sustainability, including equity, resource

---

BOX 5.1

TRENDS IN ENGINEERING SCIENCE

The current trends in engineering science can be illustrated by the trends in chemical engineering science. During many decades, situated somewhere between 1930 and 1990, chemical engineering has developed as a scientific discipline. Two paradigms can be assigned which firmly founded this trend [137]. The first one was the concept of unit operations, a principle which enables to isolate standard operations (displacement of materials, making conversions, separations) from the production process. These unit operations (e.g. pumps, distillation columns, reactors) could now be studied *in vitro*, which enabled a scientific approach. The second paradigm included physical mathematical equations in chemical engineering calculations. These physical transport phenomena proved to be very capable to apply a scientific approach to mass and heat transport processes within industrial chemical equipment. However, although both paradigms still are very powerful, chemical engineering nowadays is trying to redefine their discipline. New challenges are found in the relation consumer—processes in fabricating new products (chemical product technology), and also in green engineering [138–140]. Being heavily involved in large conversion processes closely related to resources of materials and energy, chemical engineers are able to make a significant contribution in a sustainable development [139]. In other engineering sciences, a similar trend in broadening of scope is observable.

---

depletion, climate change, biodiversity, and security of supply of energy. With 'internalization' an active usage of the knowledge domain is meant, as defined by [109,111];

2. *The competence to adequately handle tools* ('learning how to' [109] or 'methodological' [111]) which optimize designs in terms of sustainability. Tools related to the three main areas of sustainability (environmental, social, and economic) are provided in chapter 4. Two major tools relating to the whole life cycle (geometrical dimension) and to the future (temporal dimension), being LCA and scenario set building, are shortly described here.

   a. *Life cycle assessment* is the monitoring and designing of the material and energy flow, related to resources. In terms of sustainability, this is not only from an anthropocentric point of view but also from the perspective of the ecosystem [13]. Life cycle assessment therefore not only is directed to systematically rubricate the material and energy flows, but also is a way to evaluate consequences of decisions in development planning and implementation [112].

b. *Building sets of scenarios* focuses on the social and economic long-term view. Especially linked to the social and economic area, an important skill is to be able to develop a long-term view, as sustainability is strongly connected with future (generations) development [17,24,107].

LCA and building sets of scenario are the most essential and general applicable tools related to sustainability. LCA directly focuses on resources and the possibility of recycling, which is of great value of engineers, as their design usually involves large material and energy flows. Further, the key aspect of sustainability is a long-term view for which building sets of scenarios is very useful. LCA and scenario building are general applicable, irrespective of the engineering discipline, giving a good understanding of a sustainable engineering design.

3. *The competence to make well-balanced personal professional assessments* ('learning to be' or 'personal') on sustainability is connected to the capability to make responsible choices and judgments as an engineer, also in teams. The field of assessment means also dealing with uncertainties mentioned before, because one of the main features of a sustainable design is that not all requirements and specifications can be stated as fixed but shall be viewed upon as part of a dynamic process [13,24,108]. Due to the many stakeholders involved, this process of designing is making decisions, which shape the possibility and risks of products [144]. Dealing with these types of uncertainties can be looked upon as an additional competence for engineers [1].

4. *The competence to deliver sustainable and feasible design solutions* ('learning to live' or 'social'). The 'social' aspect is interpreted as that an engineer is expected to deliver a working and adequate design, within customer's constraints as specifications and costs. When the concept design is radically novel, feasible then means that the innovation pathway is included, see e.g. [4]. In the education, this means that the student also learns to apply innovation processes and that the student learns to think in a radical way, often needed to find sustainable solutions for the long term [100,113].

In the four competences, the three factors of sustainability, social, economic, and environment, all should be simultaneously addressed, depending on the design assignment [13].

These four key competences usually are additional to the common core technically oriented curriculum of engineers [108]. The difficulty for engineers is to soundly incorporate the social part, which introduces the value for the society and is connected to not-for-profit terms of the business [102]. Engineering practice usually is focused on ecological and economical aspects, as it is connected to material and energy flows, and cost engineering, respectively. Traditional engineering curricula usually do not have included these nondetermined processes as values, which are less strict, differ with time, and are even negotiable, compared to economics and ecology [102,103].

HINTS ON ACQUIRING KEY COMPETENCES

This section reports on (our) experiences of teaching of making a sustainable design. The learning on the four competences' complexity may start with reading this book, but only as a first beginning. By reading the book and exploring other relevant resources on sustainable development, it is possible to learn on concepts. However, because sustainability is much related to making assessments, it is difficult to train the basic skills on a solely individual basis and only by reading. A well-chosen case, preferably from daily practice in a work setting and connected to the work field, is a very fruitful way to directly apply the knowledge, train the connected skills, and acquire the competence to make a sustainable design.

Stakeholders in Education

The involvement of stakeholders in education, reporting from daily practice, and reflecting on the work of the students gives a more broad view on sustainability and provides a way of dealing with the aforementioned complexity. As involving different stakeholders in the design evaluation step is an important factor in making a sustainable design [112], including stakeholders in sustainable design education is advocated.

Regarding sustainability, stakeholders are broadly defined as individuals or groups who can affect the achievements of organization's objectives [79]. Representatives of stakeholders can show solutions to the complexity of sustainability issues [23,112] and testify on the necessity of applying sustainability in a social and business context. It is a way to connect theory to the real life, an important aspect in education on sustainability [114]. Including stakeholders within courses usually directly addresses normative aspects of sustainability. Experiences running up to one decade show an inspiring collaboration between students and faculty, with regional authorities and industry in regional sustainable development activities [115]. And finally, including stakeholders in an education setting adds to the participants' motivation to work on sustainability [1].

Workshop Setting

Teaching design combined with the holistic approach of sustainability requires an active form of learning which here is referred to as 'workshops' [116,117,118]. The emphasis on making an assessment, incorporating societal trends, requires interaction between participants and teachers [104,119]. An example of the usage of an active learning model is learning on ethics. As discussed in previous chapters, assessments in making a sustainable design share values and norms, which touch ethical discussions. From experience and literature on how to learn on ethics, discussing cases, on a large scale available in literature and also online [120,121], is known to be most efficient. The internal process of learners on sustainability, developing an own mindset on how they think on sustainability [122], and reflection [18] is an essential part of the learning process. Therefore, an effective form of education will be with an emphasis on learning by doing, and learning 'how to think' instead of 'what to think' [123].

This way of working resembles the well-known problem-based learning (PBL), but is less intensive. PBL learns concepts and competences around a (daily) working example or case. In this respect, PBL is found to effectively function to challenge students [117,123], but it requires a very intensive form of education [104,117]. What works is to provide a leading case throughout the courses, not as a 'leading problem', but as an illustration of the theory provided and giving the students the opportunity for a direct application of their acquired knowledge. The cases function to give feedback to the students on aspects of technical feasibility, assessment on Triple P, and the tools scenario building and application of LCA (key competence 2–4). An interactive workshop setting challenges participants to discuss assumptions and assessments of their design from which we observed that the intended learning objectives and competences are most likely to be met.

Design is Different from Scientific Research

For students trained in academic methods making a design can sometimes be very difficult to do. This is because making a design is very different from doing academic research. Showing the students the differences as presented in Table 5.1 in general helps enormously. The main hurdle to overcome is to leave the position of analyzing available information and start to generate information by designing.

In general, designers at some point in time will enter a flow of subsequent intuitive activities of synthesis, analysis, and synthesis, resulting in design solutions. It is difficult to provide knowledge on how to help this flow. Design is everywhere taught by doing [118,124,125]. The reason is that design is a combination of two ways of thinking, creative and

TABLE 5.1 Differences between Design and Science [90]

Parameter	Design	Science
Goal	Specific Solution	General Theory
Meeting goal	Good enough	Best
Method	No single method	Hypothesis and experiment
Essence	Prescriptive	Descriptive
Reliability	Calculation and judgment	Experimental validation

analytical. Because of this combination it cannot be taught by lectures or reading books only. The role of a teacher in learning on designing is also limited, but important [118]. Every human being has some creative capabilities which come to the fore by creating something new. One of the roles is creating a trust relation with the student, so that she or he feels free to try to generate preliminary solutions and is not afraid to make mistakes. A step-by-step approach creates that trust, by following the design steps problem definition, synthesis, analysis, evaluation, and reporting. Also providing a limited amount of theory, which is directly applied by the students and giving feedback on the presented intermediate results, adds to creating that trust. More practice on designing will improve the design: Lawson writes at the end of his book [90]: "It remains the case that the design process can be learned chiefly through the practice and is very difficult to teach well."

Sustainable Design is Different from Conventional Design

Engineers are increasingly involved in the future developments, not only in technical but also in nontechnical issues [1, 74]. In that respect, the role of the engineer is changing. Historically, the engineer combines capabilities of a designer with analytical reasoning from scientific disciplines [118]. Scientific inventions provided them with new tools and insights. For several decades, this mainly scientific approach appeared to be the main subject in engineers' jobs. Therefore, engineering science has got an established basis in which engineering is taught only after a solid basis in science and mathematics [118], and also simultaneously, in many cases, the design attitude of engineers has become subordinate. Box 5.1 gives a specific example for chemical engineering. However, a large share of present-day requirements is not dealing with technical issues, but includes these usually vaguely defined nontechnical issues. As technology is central in society, the experts on this, engineers, are increasingly involved in the nontechnical issues, as social and political drivers [8]. A good design of an engineer requires diversity. Diversity implies the perspectives of humanities and social sciences into design education and the practice of design [101].

In general, human beings have a natural tendency and ability to design when they have to solve a complex problem. Unfortunately, this natural tendency and ability suffer by scientific education. In scientific work, the focus is on analyzing phenomena to gain understanding. Often this understanding is then formulated as a theory describing how certain phenomena are related to other phenomena. Scientific theories are judged by being true or false. Science progresses by analyzing and criticizing published theories. This means that when students in later years have to make a design, their tendency is to treat it as a scientific problem, so they start to analyze the information given to them and if the information is little they start to do literature searches and often they criticize the information given as being incorrect and not sufficient. For these students it helps to explain the difference between designing and scientific research. Table 5.1, derived from [90] summarizing the key differences between science and design, may be helpful in this respect.

A General Setup of Courses on Making a Sustainable Design

Table 5.2 shows how a course module can be set up by applying different workshops. Table 5.3 gives an impression how workshops and stakeholders may address the four competences. Specific course elements may vary throughout the years, depending on cases and availability of stakeholders. The first and last columns of Table 5.2 are the start and finalization of the design case. The start of the case usually is experienced as being difficult, because it requires the proper definition what the aims and benefits of the design should be in terms of sustainability. At the end of the course, a forum containing representatives of society and business reflects on the participants' findings. It is recommended to include workshops on LCA, and building sets of scenarios. A separate workshop may be held on a role play in which participants experience the effect of teamwork on acquiring sustainability objectives in an industrial setting. Finally, representatives of different stakeholders, in Table 5.2 e.g., government and business, are contributing to the courses, usually connected to workshops (as LCA and scenario), obviously depending on their field of experience.

Two Examples from Practice

Finally, by describing our experiences of two courses may provide additional hints in teaching or acquiring professional assessment. The courses differ in setup, but mainly follow the contents of Table 5.2.

TABLE 5.2 Examples of Workshops in Which the Key Competences on Making a Sustainable Design are Trained

Competence	Start of Design: Problem Statement	LCA workshop	Scenario workshop	Stakeholder government	Stakeholder business	Role play	Finalization: Forum
1. Internalization of sustainability	Introduction on SD	Find key points in usage resources	Find future trends	Define role of government	Situate business	Insight in your role	Incorporate all relevant stakeholders
2. Ability to adequate handle tools (LCA and scenario building)	Define starting points for tools	Exercise	Exercise	Provide useful data	Provide useful data	Insight in teamwork	Evaluate findings of tools
3. Performing a well-balanced personal professional assessment	Make sustainability objectives concrete	Decide on technology	Select leading	Show societal trends	Apply business case (e.g. marketing)	Determine teamwork	Defend case results
4. Feasibility	Define SD	Quantify input and output flows	Test robustness of solution	Show insight in regulations	Provide technology	Give insight in team roles	Evaluate feasibility of design

TABLE 5.3   Addressing of Competences by Applying Workshops in the Learning Process and Introducing Representatives of Stakeholders

Competence	Role of stakeholder	Role of workshop in the learning process
1. Internalization of sustainability	show practical applications of implementing sustainability	provide feedback: to help students to make up their mind
2. Ability to adequate handle tools (LCA and scenario building)	help with choices in LCA and scenarios: to focus on key parameters	LCA and scenarios are typical a product of group work, build during discussions
3. Performing a well-balanced personal professional assessment	stakeholders represent professionals and the work field	workshops create conditions for debating, which is essential for a well-balanced assessment
4. Feasibility	stakeholders have access to real-life data	in presenting and discussing their results, students learn that technically sound is not always equal to usefulness

In the PhD course *Sustainable Process, Product, and System Design* (*SPPSD*) certificated students (chemical) engineering redesign an existing chemical plant, applying radical steps of improvement. The stakeholder, from industry, here acts as a problem owner, who introduces the design assignment and evaluates as a member of a forum, the students' design results. In this way, the results of the case are immediately tested and compared with practical application in an industrial setting. In the course setup, given in a full week, the morning is for lectures, while the afternoon is reserved for group work applying the theory to the course case. Feedback is organized in a finalization of each day by short group presentations of the day's findings. On Friday, each group presents the results of their casework to a panel of experts (Fig. 5.1).

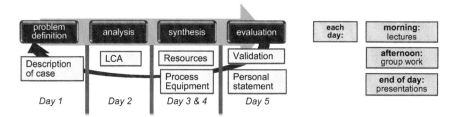

FIGURE 5.1   Outline of the PhD course on sustainability.

The course setup inherently promotes to work on a sustainable design: the focus on 'sustainable' is guaranteed by the demand of radical improvement (applying factor 4 improvement, see Box 2.1), while 'design' is directly reflected in the topics of the course days, see Fig. 5.1. The setup of the course resembles the design process [126]:

- settlement of *problem definition* for sustainable design, at the first day of the course;
- *analysis* of system elements and emission streams, which is represented by setting up an LCA at the second day;
- *synthesis* — on the third and fourth day experts provide information on specific aspects of the case, focusing on innovative sustainable technologies and solutions in context; and
- *evaluation* — the design results are presented to an industrial forum, and also participants are invited to write a personal statement in which they define their role and impact as an (chemical) engineer in a (future) work setting.

In the master course *Sustainability for Engineers* (*SfE*) students learn and practice on how to personally assess what and how they can contribute to sustainable development, depending on their position in a company or government institute. The size of the course is 5 ECTS and spread over 8 weeks. In eight meetings (half a day), each dedicated to a specific topic, (guest-) lecturers teach and coach the participants in specific aspects of sustainability. The technically oriented case is group project work and to develop and practice design and well-balanced assessment methods. A broad view on the complexity of sustainability issues is given by representatives of a variety of stakeholders, such as the government, a research institute, the business to consumer (BtC), and the business-to-business industry (BtB). Figure 5.2 gives an impression of the setup of the course.

From the courses we have learned that a workshop setting is most effective when students prepare the workshop with an assignment. For example, for each case in our course, an LCA is necessary, which is discussed in a separate workshop. As a preparation of this workshop on LCA, we shortly introduce the main concepts of LCA in an earlier meeting. The groups prepare an LCA of the case which they present at the start of the workshop. The workshop leader gives feedback on their work and adds specific information on the topic covered and on the application of sustainability in general. In this respect, participants are right from the start being involved in the workshop, which enhances the learning process.

Throughout the years, all students succeeded in attaining considerable achievements in the four competences. Evaluation from previous editions revealed that progress on the competences can be best evaluated on group work casework, combined with a personal assignment. The topic of the cases has varied throughout the years, but usually it is bound to a regional activity (own region or abroad) for which students can set

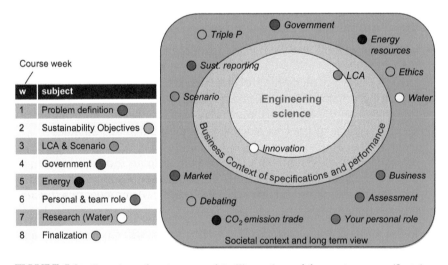

FIGURE 5.2   Overview of topics covered in (8) meetings of the master course 'Sustainability for Engineers'.

up an LCA, and sets of scenarios. Also, the students were asked to reflect on their role as an engineer, in a realistic future work setting. This included a personal statement in which they explored their own intentions on sustainability. During the courses, we observed that students were very motivated to work on the case group work and their personal assignments.

How to Apply this Book for Education

Table 5.4 describes three main forms to learn on the four competences and where in the book the information is found. In a course it is best to start with self-study by reading this book and related literature, then practice specific subjects in ½ day workshop, and finally, apply theory and practice in a course module which contains a design case. A course module may be (largely) built of workshops on specific subjects, but the design case connects them.

TABLE 5.4   Relating Four Key Competences to Learning Methods

Main competences	Self-study Concepts	Practice in workshops	Practice in course including design case
1. Internalization of sustainability	As a start: Chapter 3	Guest lectures with debate	Elaborate on specific issue
2. Making a feasible design	Chapter 2 and 4	Quick design workshop	Problem statement and evaluate results
3. Tools in making a sustainable design	Chapter 2 and 4	Specific workshops on tools	Apply tools on design case
4. Professional Assessment	Chapter 5	Role play	Personal statement, group discussions

The inspiration to be obtained from this book is closely connected to the development of the four competences. The context levels refer to different forces, ranging from planet, society to business and engineer, which all determine and influence the requirements on a sustainable design. Encouragement may be obtained by following the step-by-step design method and tools of chapters 2 and 4 and becoming aware that it works. The imagination of the context for the specific design may be enlarged by reading chapter 3. This fifth chapter providing practical proven advice on how to acquire the sustainable design competences by teaching and self-education may provide the same joy the authors obtained by writing this book.

References

[1] S.D. Sheppard, J.W. Pellegrino, B.M. Olds, On becoming a 21st century engineer, Journal of Engineering Education 97 (3) (2008) 231–234.

[2] P.M. Senge, The Necessary Revolution: How Individuals and Organizations are Working Together to Create a Sustainable World, Doubleday, New York, 2008.

[3] T.L. Friedman, Hot, Flat, and Crowded: Why We Need a Green Revolution—and How it Can Renew America, Farrar, Straus and Giroux, New York, 2008.

[4] R. Nidumolu, C.K. Prahalad, M.R. Rangaswami, Why sustainability is now the key driver of innovation, Harvard Business Review 87 (9) (2009).

[5] D.A. Lubin, D.C. Esty, The sustainability imperative, Harvard Business Review 88 (5) (2010).

[6] J. Carrillo-Hermosilla, P. Del Río, T. Könnölä, Diversity of eco-innovations: reflections from selected case studies, Journal of Cleaner Production 18 (10–11) (2010) 1073–1083.

[7] M. Braungart, W. McDonough, A. Bollinger, Cradle-to-cradle design: creating healthy emissions – a strategy for eco-effective product and system design, Journal of Cleaner Production 15 (13–14) (2007) 1337–1348.

[8] R. Adams, D. Evangelou, L. English, A.D. De Figueiredo, N. Mousoulides, A.L. Pawley, et al., Multiple perspectives on engaging future engineers, Journal of Engineering Education 100 (1) (2011) 48–88.

[9] J. Benton, Business, ethics, and the environment: imagining a sustainable future, Business Ethics Quarterly 18 (4) (2008) 567–581.

[10] World Commission on Environment and Development, Our Common Future, Oxford University Press, Oxford, New York, 1987.

[11] R. Lozano, Envisioning sustainability three-dimensionally, Journal of Cleaner Production 16 (17) (2008) 1838–1846.

[12] T. Jackson, Stockholm Environment Institute. Clean Production Strategies: Developing Preventive Environmental Management in the Industrial Economy, Lewis Publishers, Boca Raton, Fla, 1993.

[13] A.L. Carew, C.A. Mitchell, Teaching sustainability as a contested concept: capitalizing on variation in engineering educators' conceptions of environmental, social and economic sustainability, Journal of Cleaner Production 16 (1) (2008) 105–115.

[14] P.R.G. Layard, Happiness: Lessons from a New Science, Penguin Press, New York, 2005.

[15] T.H. Russ, Sustainability and Design Ethics, CRC Press, Taylor & Francis, Boca Raton, FL, 2010.

[16] R.W. Scholz, C.R. Binder, Environmental Literacy in Science and Society: From Knowledge to Decisions, Cambridge University Press, Cambridge, New York, 2011.

[17] J. Ehrenfeld, Sustainability by Design: A Subversive Strategy for Transforming our Consumer Culture, Yale University Press, New Haven, 2008.

[18] R. Morris, P. Childs, T. Hamilton, Sustainability by design: a reflection on the suitability of pedagogic practice in design and engineering courses in the teaching of sustainable design, European Journal of Engineering Education 32 (2) (2007) 135–142.

[19] J.G. Miller, Living Systems, McGraw-Hill, New York, 1978.

[20] I.M. Wallerstein, World-Systems Analysis: An Introduction, Duke University Press, Durham, 2004.

[21] T. Jackson, Prosperity without Growth: Economics for a Finite Planet, Earthscan, London, 2011.

[22] G. Unruh, Earth, Inc.: Using Nature's Rules to Build Sustainable Profits, Harvard Business Press, Boston, Mass, 2010.

[23] J. Segalàs, D. Ferrer-Balas, K.F. Mulder, What do engineering students learn in sustainability courses? The effect of the pedagogical approach, Journal of Cleaner Production 18 (3) (2010) 275–284.

[24] C.I. Davidson, H.S. Matthews, C.T. Hendrickson, M.W. Bridges, B.R. Allenby, J.C. Crittenden, et al., Adding sustainability to the engineer's toolbox: a challenge for engineering educators, Environmental Science and Technology 41 (14) (2007) 4847–4850.

[25] T.E. Graedel, B.R. Allenby, Industrial Ecology and Sustainable Engineering, Pearson, Boston, [Mass.], London, 2010.

[26] I. Serageldin, A. Steer, Making development sustainable: from concepts to action. Making Development Sustainable: From Concepts to Action 1994.

[27] S.L. Hart, Beyond greening: strategies for a sustainable world, Harvard Business Review 75 (1) (1997) 66–76.

[28] J. Elkington, Cannibals with Forks: The Triple Bottom Line of 21st Century Business, New Society Publishers, Gabriola Island, BC; Stony Creek, CT, 1998.

[29] Shell International Petroleum Company, People, Planet & Profits: The Shell Report: Summary, Shell International, London, 2001.

[30] United Nations. Department of Public Information. Johannesburg Declaration on Sustainable Development and Plan of Implementation of the World Summit on Sustainable Development: The Final Text of Agreements Negotiated by Governments at the World Summit on Sustainable Development, 26 August–4 September 2002, Johannesburg, South Africa. [S.l.]: United Nations, 2003.

[31] H. Petroski, To Engineer is Human: The Role of Failure in Successful Design, Vintage Books, New York, 1992.

[32] G. Korevaar, Sustainable Chemical Processes and Products: New Design Methodology and Design Tools (2004).

[33] K. Dorst, Understanding Design: 175 Reflections on being a Designer, Gingko Press, Corte Madera, 2007.

[34] E.U.V. Weizsäcker, A.B. Lovins, L.H. Lovins, Factor Four: Doubling Wealth, Halving Resource use: The New Report to the Club of Rome, Earthscan, London, 1997.

[35] M. Jaenicke, Ecological modernisation: new perspectives, Journal of Cleaner Production 16 (5) (2008) 557–565.

[36] S.H. Bonilla, C.M.V.B. Almeida, B.F. Giannetti, D. Huisingh, The roles of cleaner production in the sustainable development of modern societies: an introduction to this special issue, Journal of Cleaner Production 18 (1) (2010) 1–5.

[37] C. Fussler, P. James, Driving Eco-Innovation: A Breakthrough Discipline for Innovation and Sustainability, Washington DC: Pitman Pub., London, 1996.

[38] O. Gassmann, E. Enkel, H. Chesbrough, The Future of Open Innovation (2010).

[39] H.W. Chesbrough, W. Vanhaverbeke, J. West, Open Innovation: Researching a New Paradigm, Oxford University Press, Oxford, 2006.

[40] R.G. Cooper, Perspective third-generation new product processes, The Journal of Product Innovation Management 11 (1) (1994) 3–14.

[41] R.G. Cooper, S.J. Edgett, Generating Breakthrough New Product Ideas: Feeding the Innovation Funnel, Product Developemnt Institute, Ancaster, Ontario, 2007.

[42] M. Manion, Ethics, engineering, and sustainable development, IEEE Technology and Society Magazine 21 (3 SPEC.) (2002) 39–48.

[43] R.J. Batterham, Sustainability-the next chapter, Chemical Engineering Science 61 (13) (2006) 4188–4193.

[44] C.M. Fisher, A. Lovell, Business Ethics and Values: Individual, Corporate and International Perspectives, Prentice Hall/Financial Times, Harlow [etc.], 2009.

[45] D.B. Botkin, E.A. Keller, D.B. Rosenthal, Environmental Science: Earth as a Living Planet, Wiley, Hoboken, N.J, 2005.

[46] D.H. Meadows, Club of Rome. The Limits to Growth: a Report for the Club of Rome's Project on the Predicament of Mankind, Universe Books, New York, 1972.

[47] R. Devon, Towards a social ethics of engineering: the norms of engagement, Journal of Engineering Education 88 (1) (1999). 87,92+134.

[48] J.R. Mihelcic, J.B. Zimmerman, M.T. Auer, Environmental Engineering: Fundamentals, Sustainability, Design, Wiley, Hoboken, NJ, 2010.

[49] N. Myers, J. Kent, The New Atlas of Planet Management, University of California Press, Berkeley, 2005.

[50] G. Wilson, P. Furniss, R. Kimbowa, Environment, Development, and Sustainability: Perspectives and Cases from Around the World, Oxford University Press, in: association with The Open University, Oxford, New York, Milton Keynes [England], 2010.

[51] H.E. Daly, Beyond Growth: The Economics of Sustainable Development, Beacon Press, Boston, 1996.

[52] H.E. Daly, J.B. Cobb, C.W. Cobb, For the Common Good: Redirecting the Economy Toward Community, the Environment, and a Sustainable Future, Beacon Press Boston, 1989.

[53] J. Sachs, The End of Poverty: Economic Possibilities for our Time, Penguin Press, New York, 2005.

[54] T. Princen, The Logic of Sufficiency, MIT Press, Cambridge, MA, 2005.

[55] L.R. Brown, L.R. Brown, Earth Policy Institute. Plan B 3.0: Mobilizing to Save Civilization, W.W. Norton, New York, 2008.

[56] A. Gore, Melcher Media, An Inconvenient Truth: The Planetary Emergency of Global Warming and What We can do about it, Rodale Press, New York, 2006.

[57] D. Nierenberg, B. Halweil, L. Starke, Worldwatch Institute, State of the World 2011: Innovations that Nourish the Planet: A Worldwatch Institute Report on Progress Toward a Sustainable Society, W.W. Norton & Co, New York, 2011.

[58] E. Assadourian, L. Starke, L. Mastny, Worldwatch Institute, State of the world, 2010: Transforming Cultures: From Consumerism to Sustainability: A Worldwatch Institute Report on Progress Toward a Sustainable Society, W.W. Norton, New York, NY, 2010.

[59] L. Starke, Worldwatch Institute, State of the world 2009: Into a Warming World: A WorldWatch Institute Report on Progress Toward a Sustainable Society, W.W. Norton & Co, New York, 2009.

[60] J. Rockström, W. Steffen, K. Noone, Å Persson, F.S. Chapin, E.F. Lambin, A safe operating space for humanity, Nature 461 (7263) (2009) 472–475.

[61] Millennium Ecosystem Assessment (Program), Our Human Planet: Summary for Decision-Makers, Island Press, Washington [D.C.], 2005.

[62] Millennium Ecosystem Assessment (Program), Millennium Ecosystem Assessment (2001).

[63] M. Pollan, The Omnivore's Dilemma: A Natural History of Four Meals, Penguin Press, New York, 2006.

[64] D. Archer, The Global Carbon Cycle, Princeton University Press, Princeton, 2010.

[65] M. Herva, A. Franco, E.F. Carrasco, E. Roca, Review of corporate environmental indicators, Journal of Cleaner Production, 2011.

[66] International Energy Agency, OECD, Key World Energy Statistics 2010 (2010).

[67] Organisation for Economic Co-operation and Development, Agency IE, World Energy Outlook 2009 (2009).

[68] International Energy Agency, Organisation for economic co-operation and development, source OECD (online service), Tracking Industrial Energy Efficiency and CO_2 Emissions in Support of the G8 Plan of Action (2007).

[69] World Business Council on Sustainable Development. Vision 2050. Earthprint.com., 2010.

[70] United Nations Development Programme, Human Development Report 2009: Overcoming Barriers: Human Mobility and Development, Palgrave Macmillan, New York, Basingstoke: United Nations, 2009.

[71] G.T. Gardner, T. Prugh, L. Starke, Worldwatch Institute, State of the World 2008: Innovations for a Sustainable Economy: A Worldwatch Institute Report on Progress Toward a Sustainable Society, W.W. Norton, New York, London, 2008.

[72] E-atlas Worldbank Millennium Goals.

[73] Worldbank Millennium Goals.

[74] National Academy of Engineering, Grand Challenges for Engineering (2008).

[75] C.M. Vest, Context and challenge for twenty-first century engineering education, Journal of Engineering Education 97 (3) (2008) 235–238.

[76] K. Haghighi, K.A. Smith, B.M. Olds, N. Fortenberry, S. Bond, The time is now: are we ready for our role? Journal of Engineering Education 97 (2) (2008) 119–121.

[77] J.P. Holdren, Presidential address: science and technology for sustainable well-being, Science 319 (5862) (2008) 424–434.

[78] D. Nierenberg, Worldwatch Institute, State of the World 2006: A Worldwatch Institute Report on Progress Toward a Sustainable Society, W.W. Norton, New York, London, 2006.

[79] L.A. Perez-Batres, V.V. Miller, M.J. Pisani, Institutionalizing sustainability: an empirical study of corporate registration and commitment to the United Nations global compact guidelines, Journal of Cleaner Production 19 (8) (2011) 843–851.

[80] M.E. Porter, M.R. Kramer, Strategy & society: the link between competitive advantage and corporate social responsibility, Harvard Business Review 84 (12) (2006) 78–92.

[81] G. Unruh, R. Ettenson, Growing green, Harvard Business Review 88 (6) (2010).

[82] S.L. Hart, M.B. Milstein, J. Caggiano, Creating sustainable value, Academy of Management Executive 17 (2) (2003) 56–69.

[83] S. Robinson, Engineering, Business and Professional Ethics, Elsevier/Butterworth-Heinemann, Amsterdam, Boston, 2007.

[84] A. Van Der Heijden, P.P.J. Driessen, J.M. Cramer, Making sense of corporate social responsibility: exploring organizational processes and strategies, Journal of Cleaner Production 18 (18) (2010) 1787–1796.

[85] Global Compact Office, UN Global Compact Ten Principles. Annual Review/UN Global Compact Office (2010).

[86] T. Dyllick, K. Hockerts, Beyond the business case for corporate sustainability, Business Strategy and the Environment 11 (2) (2002) 130–141.

[87] J.F. Molina-Azorín, E. Claver-Cortés, M.D. López-Gamero, J.J. Tarí, Green management and financial performance: a literature review, Management Decision 47 (7) (2009) 1080–1100.

[88] A. Colby, W.M. Sullivan, Ethics teaching in undergraduate engineering education, Journal of Engineering Education 97 (3) (2008) 327–338.

[89] B.M. Beamon, Environmental and sustainability ethics in supply chain management, Science and Engineering Ethics 11 (2) (2005) 221–234.

[90] B. Lawson, How Designers Think: The Design Process Demystified, Elsevier/Architectural, Oxford, Burlington, MA, 2006.

[91] V. Pastoor, De Ingenieur 16 (2011) 28–29.

[92] M. Thomas, D. Brain, Crowd Surfing: Surviving and Thriving in the Age of Consumer Empowerment, A. & C. Black, London, 2008.

[93] G.J. Harmsen, L.A. Chewter, Industrial applications of multi-functional, multi-phase reactors, Chemical Engineering Science 54 (10) (1999) 1541–1545.

[94] G.J. Harmsen, Industrial best practices of conceptual process design, Chemical Engineering and Processing: Process Intensification 43 (5) (2004) 677–681.

[95] G.J. Harmsen, Reactive distillation: the front-runner of industrial process intensification. A full review of commercial applications, research, scale-up, design and operation, Chemical Engineering and Processing: Process Intensification 46 (9) (2007) 774–780.

[96] A. Niederl-Schmidinger, M. Narodoslawsky, Life cycle assessment as an engineer's tool? Journal of Cleaner Production 16 (2) (2008) 245–252.

[97] M. Hauschild, H. Wenzel, Environmental Assessment of Products, vol. 2, Chapman & Hall, London, 1998. Scientific Background.

[98] L. Alting, M. Hauschild, H. Wenzel, Environmental Assessment of Products. 1, Methodology, Tools and Case Studies in Product Development, Chapman & Hall, London [u.a.], 2000.

[99] J.M. Douglas, Conceptual Design of Chemical Processes, McGraw-Hill, New York, 1988.

[100] C.M. Bacon, D. Mulvaney, T.B. Ball, E.M. DuPuis, S.R. Gliessman, R.D. Lipschutz, et al., The creation of an integrated sustainability curriculum and student praxis projects, International Journal of Sustainability in Higher Education 12 (2) (2011) 193–208.

[101] C.L. Dym, J.W. Wesner, L. Winner, Social dimensions of engineering design: observations from mudd design workshop III, Journal of Engineering Education 92 (1) (2003) 105–107.

[102] L.E. Schlange, Stakeholder identification in sustainability entrepreneurship, Greener management international (55) (2009) 13–32.

[103] A. El-Zein, D. Airey, P. Bowden, H. Clarkeburn, Sustainability and ethics as decision-making paradigms in engineering curricula, International Journal of Sustainability in Higher Education 9 (2) (2008) 170–182.

[104] G. Steiner, A. Posch, Higher education for sustainability by means of transdisciplinary case studies: an innovative approach for solving complex, real-world problems, Journal of Cleaner Production 14 (9–11) (2006) 877–890.

[105] J. Davtdh, D. Shen, R.M. Marra, C.H.O. Young-Hoan, L.L.O. Jenny, V.K. Lohani, Engaging and supporting problem solving in engineering ethics, Journal of Engineering Education 98 (3) (2009) 235–254.

[106] F.J. Lozano García, K. Kevany, D. Huisingh, Sustainability in higher education: what is happening? Journal of Cleaner Production 14 (9–11) (2006) 757–760.

[107] M. Barth, J. Godemann, M. Rieckmann, U. Stoltenberg, Developing key competencies for sustainable development in higher education, International Journal of Sustainability in Higher Education 8 (4) (2007) 416–430.

[108] Y. Mochizuki, Z. Fadeeva, Competences for sustainable development and sustainability: significance and challenges for ESD, International Journal of Sustainability in Higher Education 11 (4) (2010) 391–403.

[109] J. Segalàs, D. Ferrer-Balas, M. Svanström, U. Lundqvist, K.F. Mulder, What has to be learnt for sustainability? A comparison of bachelor engineering education competences at three european universities, Sustainability Science 4 (1) (2009) 17–27.

[110] S. Geertshuis, Improving decision making for sustainability: a case study from New Zealand, International Journal of Sustainability in Higher Education 10 (4) (2009) 379–389.

[111] J. Delors, Unesco, Learning: The Treasure within: Report to UNESCO of the International Commission on Education for the Twenty-First Century, UNESCO Pub., Paris, 1998.

[112] L. Thabrew, A. Wiek, R. Ries, Environmental decision making in multi-stakeholder contexts: applicability of life cycle thinking in development planning and implementation, Journal of Cleaner Production 17 (1) (2009) 67–76.

[113] D. Millet, L. Bistagnino, C. Lanzavecchia, R. Camous, T. Poldma, Does the potential of the use of LCA match the design team needs? Journal of Cleaner Production 15 (4) (2007) 335–346.

[114] K. Brundiers, A. Wiek, C.L. Redman, Real-world learning opportunities in sustainability: from classroom into the real world, International Journal of Sustainability in Higher Education 11 (4) (2010) 308–324.

[115] G. Zilahy, D. Huisingh, The roles of academia in regional sustainability initiatives, Journal of Cleaner Production 17 (12) (2009) 1057–1066.

[116] H. von Blottnitz, Promoting active learning in sustainable development: experiences from a 4th year chemical engineering course, Journal of Cleaner Production 14 (9–11) (2006) 916–923.

[117] K.F. Mulder, J. Segalas-Coral, D. Ferrer-Balas, Educating engineers for/in sustainable development? What we knew, what we learned, and what we should learn, Thermal Science 14 (3) (2010) 625–639.

[118] C.L. Dym, A.M. Agogino, O. Eris, D.D. Frey, L.J. Leifer, Engineering design thinking, teaching, and learning, Journal of Engineering Education 94 (1) (2005) 103–119.

[119] P. Hopkinson, P. James, Practical pedagogy for embedding ESD in science, technology, engineering and mathematics curricula, International Journal of Sustainability in Higher Education 11 (4) (2010) 365–379.

[120] G. Surie, A. Ashley, Integrating pragmatism and ethics in entrepreneurial leadership for sustainable value creation, Journal of Business Ethics 81 (1) (2008) 235–246.

[121] M. Gorman, M. Hertz, G. Louis, L. Magpili, M. Mauss, M. Mehalik, et al., Integrating ethics & engineering: a graduate option in systems engineering, ethics, and technology studies, Journal of Engineering Education 89 (4) (2000). 461, 469+505–508.

[122] D.M. Podger, E. Mustakova-Possardt, A. Reid, A whole-person approach to educating for sustainability, International Journal of Sustainability in Higher Education 11 (4) (2010) 339–352.

[123] I. Thomas, Critical thinking, transformative learning, sustainable education, and problem-based learning in universities, Journal of Transformative Education 7 (3) (2009) 245–264.

[124] S. Purzer, J. Chen, Teaching decision-making in engineering: A review of textbooks and teaching approaches, in: 2010.

[125] C.J. Atman, M.E. Cardella, J. Turns, R. Adams, Comparing freshman and senior engineering design processes: an in-depth follow-up study, Design Studies 26 (4) (2005) 325–357.

[126] G.J. Harmsen, H.H. Kleizen, G. Korevaar, S.M. Lemkowitz, Chemical technology sustainable design course: key success elements for creativity, in: EESD 2004: Engineering Education in Sustainable Development, Barcelona.

[127] B. Jastorff, R. Störmann, U. Wölcke, Struktur-Wirkungs-Denken in Der Chemie: Eine Chance für Mehr Nachhaltigkeit, Aschenbeck & Isensee, Bremen [u.a.], 2003.

[128] G. Boyle, Open University. Renewable Energy: Power for a Sustainable Future, Oxford University Press in Association with the Open University, Oxford, 1996.

[129] A. Azapagic, S. Perdan, R. Clift, Sustainable Development in Practice: Case Studies for Engineers and Scientists, Wiley, Hoboken, NJ, 2004.

[130] B. Beloff, M. Lines, D. Tanzil, Transforming sustainability strategy into action: the chemical industry, Wiley-Interscience, Hoboken, N.J, 2005.

[131] C. Holliday, Sustainable growth, the DuPont way, Harvard Business Review 79 (8) (2001). 129, 134, 162.

[132] R.J. Orsato, P. Wells, The automobile industry & sustainability, Journal of Cleaner Production 15 (11−12) (2007) 989−993.

[133] P. Anastas, J.C. Warner, Green Chemistry: Theory and Practice, Oxford University Press, Oxford, 2000.

[134] J.J. Siirola, Industrial applications of chemical process synthesis, Advances in Chemical Engineering 23 (C) (1996) 1−62.

[135] A.I. Stankiewicz, J.A. Moulijn, Re-Engineering the Chemical Processing Plant: Process Intensification, M. Dekker, New York, 2004.

[136] G. Scholes, Integrating Sustainable Development into the Shell Chemicals Business, in: W.P.M. Weijnen, P.M. Herder (Eds.), Environmental Performance Indicators in Process Design and Operation, Delft University Press, 1999, p. 69.

[137] E. Favre, V. Falk, C. Roizard, E. Schaer, Trends in chemical engineering education: process, product and sustainable chemical engineering challenges, Education for Chemical Engineers 3 (1) (2008) e22−e27.

[138] J. García-Serna, L. Pérez-Barrigón, M.J. Cocero, New trends for design towards sustainability in chemical engineering: green engineering, Chemical Engineering Journal 133 (1−3) (2007) 7−30.

[139] R. Costa, G.D. Moggridge, P.M. Saraiva, Chemical product engineering: an emerging paradigm within chemical engineering, AICHE Journal 52 (6) (2006) 1976−1986.

[140] R.M. Voncken, A.A. Broekhuis, H.J. Heeres, G.H. Jonker, The many facets of product technology, Chemical Engineering Research and Design 82 (11) (2004). 1411−1124.

[141] J. Harmsen, J.B. Powell, Sustainable development in the process industries, Wiley, Oxford, 2010.

[142] Source of Figures 3.5 and 3.6: Sustainable Development Commission, Report "Prosperity to Growth", March 2009, www.sd-commission.org.uk, accessed December 2011.

[143] A.I. Stankiewicz, J.A. Moulijn, Re-engineering the chemical processing plant : process intensification. M. Dekker, New York, 2004. Table 3, Generally Recognized Environmental Impacts.

[144] A. van Gorp, Ethical issues in engineering design processes; regulative frameworks for safety and sustainability. Des Stud (2) (2007) 117−131.

Index

Note: Page numbers followed by f indicate figures, t indicate tables and b indicate boxes.

Printed and bound by CPI Group (UK) Ltd, Croydon, CR0 4YY

03/10/2024

01040419-0009